Nature and the Environment in Amish Life

Young Center Books
in Anabaptist and Pietist Studies
Steven M. Nolt, *Series Editor*

Nature and the Environment in Amish Life

David L. McConnell and Marilyn D. Loveless

JOHNS HOPKINS UNIVERSITY PRESS Baltimore

 Published 2018
Printed in the United States of America on acid-free paper
9 8 7 6 5 4 3 2

Johns Hopkins University Press
2715 North Charles Street
Baltimore, Maryland 21218-4363
www.press.jhu.edu

Library of Congress Cataloging-in-Publication Data

Names: McConnell, David L., 1959–, author. | Loveless, Marilyn D., author.
Title: Nature and the environment in Amish life / David L. McConnell and Marilyn D. Loveless.
Description: Baltimore : Johns Hopkins University Press, 2018. | Series: Young Center books in Anabaptist and Pietist studies | Includes bibliographical references and index.
Identifiers: LCCN 2017058444 | ISBN 9781421426167 (pbk. : alk. paper) | ISBN 9781421426174 (electronic) | ISBN 1421426161 (pbk. : alk. paper) | ISBN 142142617X (electronic)
Subjects: LCSH: Human ecology—Religious aspects—Amish. | Human ecology—United States. | Amish—Customs and practices.
Classification: LCC GF80 .M373 2018 | DDC 261.8/808828973—dc23
LC record available at https://lccn.loc.gov/2017058444

A catalog record for this book is available from the British Library.

Special discounts are available for bulk purchases of this book. For more information, please contact Special Sales at 410-516-6936 or specialsales@press.jhu.edu.

Johns Hopkins University Press uses environmentally friendly book materials, including recycled text paper that is composed of at least 30 percent post-consumer waste, whenever possible.

CONTENTS

At a time when more than three-quarters of the nation's population live in cities, Americans are paradoxically flooded with messages about the importance of returning to nature. We hear that immersing ourselves in gardening, hiking, and other outdoor activities can improve our mental and physical health. Encouraging our children to unplug from the electronic world will prevent them from contracting the psychological malady Richard Louv calls "nature deficit disorder."[1] Being attentive to the plight of our local and our global natural environment will help us reverse the increasing strain that humans are placing on ecological systems. If individuals, households, corporations, and governments don't reduce consumption and waste, we are told, the very future of the planet will be in jeopardy. It's enough to make one ask, as one blogger on sustainability did, "Why couldn't we all live like the Amish? If we all lived like they do, this host of modern problems we have would disappear."[2]

The chances that Americans will embrace the Amish as a model for cultural and environmental renewal are slim to none—and for good reason. The Amish world-view does not provide a viable blueprint for living in a pluralistic world and therefore, as the sociologist Marc Olshan reminds us, is "not an option for most individuals or for us as a society."[3] But we believe that the assumptions underlying the blogger's question—that the Amish live a simple existence, close to the land, and illustrate a more sustainable and earth-friendly lifestyle—are worth further consideration. The example of a seemingly preindustrial people living in the midst of a modern society offers a useful mirror for examining our changing relation with the natural world.

Over the past two decades, many of the popular myths about the Amish—that they are slowly dying out, that they are a homogenous group of technophobes, that they are "stuck in the past"—have been convincingly dispelled.[4] Yet the image of the Amish as living in harmony with nature is alive and well. Today's Amish brand, whether attached to furniture, quilts, or fresh produce, has become synonymous with a rural, wholesome, and authentic lifestyle. Moreover, because outsiders view the Amish as closely tied to the land, distancing themselves from modern conveniences, they naturally assume that the Amish lead and embrace an environmentally minded lifestyle.

But is this view consistent with how the Amish themselves see their engagements with the natural world, or are outsiders misinterpreting Amish life on the basis of superficial interactions? Do Amish lifestyle choices and attitudes toward nature offer new ways of thinking and behaving that can inform our individual and collective efforts to tread more lightly on the earth? In this book, we explore the image of the all-natural Amish from a diverse set of viewpoints within and outside the Amish community. While there have been autobiographical accounts by outsiders who temporarily adopted the "simplicity" of the Amish way of life, to our knowledge this is the first in-depth, scholarly account of how the image of the environmentally conscious Amish fits with the changing realities of Amish lives.

We pose two central questions. First, we ask how the Amish understand and use nature in their daily lives. Framing the question in such a broad way highlights our interest in how the Amish explain and interpret their surroundings, what nature provides for their well-being, and the behavioral choices they make at home, at work, and at play. Second, we ask how Amish practices impact the natural environment. We explore the ways that Amish beliefs and choices intersect with and potentially inform non-Amish understandings of ecology, sustainability, and environmentally sound lifestyles. We ask how the practices of the Amish compare with those of their rural non-Amish neighbors and how they are perceived by non-Amish professionals with whom they work (extension agents, foresters, veterinarians, and others). We also explore the complicated ways in which diverse Amish views and practices both align with and run counter to science-based environmentalism.

In seeking to answer these questions, we hope to contribute in a small way to several broader conversations about the relation between religion, culture, and use of the earth's natural resources. "Sustainability" has become an increasingly prevalent, if poorly defined, catchphrase for many kinds of human endeavors involving resource use.[5] Populations of agricultural smallholders who retain a connection to the land are often assumed to be practicing sustainable behaviors simply by default. Such groups do not live in a vacuum, though. Many Amish have left farming behind, adopting new types of consumptive and extractive nature-based businesses and leisure activities that complicate our understanding of stewardship of the earth. As a fast-growing population with deep ties to the land who have also enjoyed remarkable success as small-business entrepreneurs, the Amish are well positioned to help us reflect on the claim that "a market economy, population increase, and the new technologies of capitalism are inevitably at odds with sustainable systems."[6]

Second, we hope to contribute to the much-debated question of whether certain religious traditions promote world-views that are relatively respectful of the earth's resources.[7] The anthropologist Leslie Sponsel argues that the image of indigenous peoples "using spiritual doctrine to protect dwindling natural resources . . . has proven attractive to Western audiences" because it seems to "hold up a critical mirror to the secular character of resource-degrading life in industrialized societies."[8] But Sponsel notes that such a view may confuse the consequences of religious doctrine with a purposive commitment to conservation.[9] The Amish present an interesting case in this regard because choices that seem to minimize ecological harm may be more a by-product of their religious beliefs than a reflection of ecological-mindedness. Their example throws into sharp relief the relation between intentionality and outcome that applies more generally to all conservation efforts.

Our final goal in conducting this study is to contribute in a small but meaningful way to the increasingly global conversation about the fate of the earth and its resources. It is hard to miss the sobering reports about the effects of human activities on the earth, including accelerated rates of species extinction, higher levels of carbon in the atmosphere, diminishing reservoirs of clean water on the planet, and concerns about whether the world's food supply can keep up with global population growth.[10] The fact that human activity has left "a pervasive and persistent signature on the Earth itself" has led many scientists to argue for a new geological time unit, the Anthropocene, to distinguish this era from the Holocene.[11] Some of these accounts of planetary changes are starkly apocalyptic, while others express optimism that human ingenuity will prevail in the long run. We share this concern about the world's current trajectory and ask what the appeal of the pastoral image of the Amish tells us about our own misgivings, trepidations, and anxieties surrounding the impacts our lives are having on the planet.

Our journey to understand Amish views of nature began more than fifteen years ago when coauthor McConnell, an anthropologist, was working on his previous book, *An Amish Paradox*. Through a series of introductions, he had agreed to pick up six Amish acquaintances at 4:00 a.m. near Mt. Hope, Ohio, for an all-day birding trip along the Lake Erie shoreline. As a moderately serious birder, he thought he was prepared, but as daylight came and his passengers piled out of the van at the first stop, he soon realized that he was completely overmatched. His inexpensive pair of binoculars paled in comparison with the thousands of dollars' worth of optical equipment, including state-of-the-art Swarovski spotting scopes, that his birding companions had brought with them. And this outing was no leisurely stroll in the park. His Amish companions were extremely skilled at spotting and identifying birds, and they birded at a torrid pace. They were keenly attuned to habitat, songs and calls, behavior, and other environmental cues of bird life. Most kept life lists, year lists, county lists, state lists, and yard lists of all the birds they had seen. By the end of the day he was exhausted and exhilarated, but he realized that he had stumbled on a topic worth further exploration.

Fast-forward ten years. *An Amish Paradox* was finished, and it was time to choose a new project. In the meantime, McConnell had agreed to be the driver for an

Amish family (parents and three teenage children) on a birding vacation in the southwestern United States. Coauthor Loveless, a College of Wooster colleague, plant ecologist, and also a birder, was living in Tucson that summer, and she invited everyone to stay at her house for a week while visiting local points of interest. This was her first in-depth interaction with the Amish, and she hit it off extremely well with the family. As we visited Cave Creek Canyon, the Arizona–Sonora Desert Museum, and other natural sites of interest, she too became intrigued by how the Amish viewed and interacted with natural landscapes. A year later we decided to collaborate on this project on Amish views of nature.

Ordinarily, negotiating access to Amish communities would be very difficult, but because of contacts made during previous research, we were able to draw on a broad network of Amish friends and acquaintances for informal conversations and formal interviews. All told, we interviewed more than 150 individuals, including Amish from thirty-five different settlements and fifteen different affiliations across twelve states. We also interviewed several dozen non-Amish individuals who work closely with the Amish. Occasionally, we identify Amish or non-Amish individuals who have used their own names in published works, but in all other cases we honor the request for anonymity by those with whom we spoke. In addition to conducting interviews, we attended many relevant public events, visited Amish settlements in Colorado, Wyoming, and Montana, conducted a questionnaire survey that formed the basis of our discussion of the Amish ecological footprint, and collected a mountain of Amish-authored newsletters, brochures, magazines, and books. For readers who want to know more, we further describe our methodological approach, along with its strengths and weaknesses, in the appendix.

On numerous occasions during our fieldwork, we were reminded of the saying "A bird is not an ornithologist." Just because you *are* something doesn't mean you can explain it. Some Amish have been puzzled by what we are trying to study. Because they are used to relating to nature on an experiential level, it's hard to understand a study that approaches nature in a more abstract way. We asked questions that Amish often don't ask, trying to probe assumptions that, as one New Order man put it, "are part of the air." Where possible, we tried to use this problem of cultural translation as a window into Amish thinking and as a critical lens on our own assumptions. Though we conducted all of our interviews in English, we learned that there is no precise Pennsylvania Dutch equivalent for the English word "environment." Even the term "nature" (*natur*) is more often used to refer to *human* nature than to the physical environment, or as one Amish man put it, to "something that's inescapable. It's ordained by God." The Amish use the term *landschaft* to refer to the landscape in a region, but other terms, such as *Schopfung* (Creation) or *veld* (world) carry very specific positive or negative moral connotations. Recognizing that our study might seem distant from everyday experience, we tried to keep our questions grounded in specifics even as we sought for deeper reflection.

We are keenly aware that our personal backgrounds and perspectives as social and natural scientists lead us to think about nature somewhat differently than it is approached in the Plain community, where, to quote the *Gemeinde Register*, "apart from divine providence and power, nature is absolutely nothing."[12] We have tried to set aside our personal beliefs in order to write a fair and balanced account, but it is impossible for researchers not to be affected by the people they meet during fieldwork. We lost count of the number of times a comment or an observation by our Amish hosts would prompt us to reflect on our own assumptions about science, religion, and environmentalism. For that, we are extremely grateful, and we hope that our book spurs a broader conversation between Amish and non-Amish—in keeping with one minister's sentiment that "many Amish would like to see how we compare on the radar of the rest of the world."

To provide organizational structure for our wide-ranging inquiry, we lay out the research puzzle and its background in the first chapter. The remainder of the book is divided into four parts that examine Amish interfaces with nature: in the home, at work, during leisure time, and in the context of environmental policies in the wider society. In part I, "Growing Up Rural," we explore the role of nature and science in Amish families and schools, and we compare the carbon footprint of Amish households with that of a comparable rural, non-Amish population. Part II, "Working with Nature," examines new forms of nature-based livelihoods in agriculture, forestry, and animal husbandry through the lens of ecological sensitivity and environmental impact. In part III, "Reconfiguring the Outdoors," we ask how Amish appreciate and value nature as a source of pleasure and leisure, from gardening and herbal remedies to outdoor recreation, travel, and nature writing. Finally, in part IV, "The Amish as Environmentalists," we consider how Amish ways of life intersect with environmental issues that are manifest both within their local or regional community and in the larger world.

We employ several stylistic conventions that follow Kraybill, Johnson-Weiner, and Nolt's definitive work on the Amish. We consider as Amish any group that affirms the basic tenets in the 1632 Dordrecht Confession of Faith, such as adult baptism and separation from the world; uses horse-drawn transportation; speaks a German-derived dialect; and considers itself Amish.[13] We also use the terms "non-Amish," "English," and "outsiders" interchangeably to refer to people who are outside Amish society. We have tried to be as careful as possible when using the terms "liberal/progressive" and "conservative" (and "high" and "low") to refer to Amish groups that have made different compromises with modernity. But we rely on the reader's recognition that these generalizations almost always include exceptions. Finally, we use the term "church district" to refer to local congregations, usually twenty to forty families, that have their own Ordnung, an unwritten set of prescriptions and proscriptions that govern the conduct of church members, overseen by several male ordained leaders, usually a bishop, two ministers, and a deacon. In contrast, "affiliation" describes a theologically similar cluster of church districts that consider themselves "in fellowship" with one another, while "settlement" refers to "church districts that share a common history in a given geographical area."[14]

It is not our intent to discredit the Amish in any way in this book. Our perspective from the outset has been that of interested outsiders who became curious about Amish views of nature and whose journey to understand how the Amish construct the human-nature interface yielded some surprising and complicated results. When measured against an environmental yardstick, the Amish sometimes fall short on certain issues, but so do most people who live in industrialized nations, including ourselves. In other cases, the Amish behave in an environmentally responsible manner, but not always for ecological reasons. More often than not, the answer to the question "How ecological are the Amish?" is far from straightforward. It depends on which group of Amish, which issue, which time frame, which region of the country.

We strongly believe, however, that holding on to a simplistic image of the Amish as environmentally conscious or not demeans their real humanity.[15] It also sets them up for charges of hypocrisy, such as the comment one Old Order woman told us she heard as she carried a jug of milk out of Walmart: "Don't you Amish have your own cows at home?" The seductive appeal of a horse-and-buggy society living close to the land and in total harmony with nature actually forestalls meaningful conversation about the impacts of modernity on Amish life. We try to provide a balanced appraisal that humanizes the Amish, pointing out both distinctive aspects of their lifestyle and the many similarities they share with rural non-Amish residents. Idealizing Amish society does a great disservice to all who wish to understand the Amish as ordinary people with strengths and weaknesses like anyone else.

ACKNOWLEDGMENTS

Over the seven years of fieldwork for this book, we received help from many people and institutions that made our efforts go much more smoothly. First and foremost, to the many Amish and non-Amish individuals and families who talked with us, traveled with us, and responded to our questions, we express our deepest gratitude. You tolerated our incessant questions, you shared your perspectives about your community, and you continually delighted us with your insights. You helped us to identify new sources of information, and you provided guidance when we had misconceptions. We were honored by your generosity and your good humor, and we hope this book is an accurate and careful reflection of the thoughts you shared with us.

For our visits to Amish settlements throughout the Midwest and the western United States, many Amish individuals introduced us to relatives in those communities, while others took time out of their schedules to guide and accompany us. We are exceedingly grateful for the generosity of spirit that we encountered at every turn. Several non-Amish friends and acquaintances were also invaluable in helping us to gain access to Amish communities outside Ohio: Mike Carey, Saloma Furlong, Erik Wesner.

At various points in our project, many colleagues near and far assisted us in clarifying our thinking and suggested directions we might consider: Leslie Sponsel, Bron Taylor, Richard Moore, Richard Stevick, Susan Clayton, Daniel Bourne, Matthew Mariola, Caroline Brock, Amyaz Moledina, David Wiesenberg, Susan Lehman, Mark Weaver, J. Douglas Drushal. Erik Wesner invited us to compose blog posts for his Amish America website, which helped us refine some of our thinking and provided useful comments from readers. Joseph Donnermeyer made available his database on Amish settlements to create figure 1, and Shelley Judge and Ric Reynolds helped us turn those data into an actual map. A special thanks to student research assistants at the College of Wooster who put in countless hours transcribing interviews, especially Christopher Perrin but also Clare Carlson, Sidney Irias, Paris Nahas, Myra Prami, Haley Skeens, and Tiffany Trunk.

As we drafted and sent out the Household Resource Use Survey, we benefited from suggestions by Richard Lehtinen and Matthew Mariola and by various Amish friends about how to best phrase certain questions. Virginia Pett vetted our calculations for modifying carbon footprints and pointed out places where we needed to be more transparent. Amber Garcia was especially helpful in walking us through statistical analyses for the New Ecological Paradigm survey and ensuring that we had appropriately summarized our results. Joyce Heitger was efficient and helpful in assisting us with creating and mailing the surveys. We thank all these colleagues and friends for their willing input.

Our project would never have gotten off the ground without the encouragement of Donald Kraybill, whose guidance and support were crucial in the early stages. We also owe a huge debt of gratitude to Steven Nolt, who provided timely, thoughtful, and constructive feedback on drafts of every chapter. Both the substantive arguments and the supporting details are much improved because of his suggestions. The manuscript also benefited from very helpful comments by two anonymous reviewers. Three Amish reviewers read and commented on the entire manuscript, and we have attempted to use their advice to make the book stronger and more readable. Many people deserve credit for their help on our project, but any errors or misinterpretations in the text are ours, alone.

Our research was made possible by grants from the Great Lakes College Association (GLCA) New Directions Program, the Henry Luce III Fund for Distinguished Scholarship, the College of Wooster Faculty Development Fund, and the College of Wooster Research Leaves Program. We appreciate Heather Fitz Gibbon's assistance in helping us navigate the complicated terrain of these funding opportunities.

Finally, we owe our deepest thanks to members of our respective families. Cathy McConnell provided excellent suggestions on numerous drafts of chapters and was of great assistance in the logistics of sending out our survey. She accompanied us on a number of fieldwork-related outings over the years and provided intangible support in countless ways. Owen and Pat McConnell provided insightful suggestions from their perspectives as environmentalists and naturalists, while Brennen McConnell, Alaina and Nick Nutile, and Jack and Jeannine Love provided sympathetic ears and good humor when needed. The Loveless siblings and their partners, John and Kathy, Jan and Bill, and Steve and Ann, provided steady support for this endeavor by asking timely questions and listening to long discourses, both of which helped to clarify our thinking. Other family members and friends provided moral support in many forms throughout the process. Nan Uhl was an indispensable source of encouragement and clarity when it was most needed.

We hope that both non-Amish and Amish readers will find information in this book that is interesting and useful and that they will be prompted, as we were on innumerable occasions, to reflect on their own engagement with nature and the environment that surrounds us.

Nature and the Environment in Amish Life

Deciphering the Amish Relationship with Nature

In the late 1990s Matthew and Nancy Sleeth were living the American dream in a small New England town, enjoying the affluent family life that came with Matthew's successful career as a physician. But a series of family crises led them to realize that their current life was neither sustainable nor spiritually meaningful. Worried that the earth was dying and with a newfound determination to do their part to change things, Matthew quit his job and the entire family moved to Kentucky. Once there, they established a faith-based environmental nonprofit organization, Blessed Earth, which they co-direct today. In the process, the Sleeths dramatically downsized their own ecological footprint, giving away half their possessions and moving to a house the size of their old garage. They began holding streambed cleanups, organizing tree-planting efforts, and coordinating discussions across the nation on faith and ecology. Today this "evangelical Christian family who hug trees"[1] has become a well-known voice in what is called the Creation Care movement.

What inspired the Sleeths to so transform their way of life? A key source of their inspiration is captured in the title of Nancy Sleeth's 2009 book, *Almost Amish: One Woman's Quest for a Slower, Simpler, More Sustainable Life*. In this highly readable paperback, Sleeth recounts the family's ecological and spiritual reawakening based on what she identifies as ten fundamental Amish virtues that provide a recipe for getting back to basics. The Amish values Sleeth identifies include keeping homes simple and uncluttered, building community, shopping locally, and cultivating family ties. Sleeth also extols the value of reconnecting with nature, since "time spent in God's creation reveals the face of God."[2] To recalibrate one's relationship with nature, "screen time" must be replaced with "green time," such as playing outdoors, growing a garden, or planting a tree. Though Sleeth admits that the Amish are not perfect, her book unabashedly holds up their overall lifestyle as an object of admiration and emulation and as a means for the wider society to address environmental crises.

About the same time that the Sleeths were rethinking their family's ecological priorities in light of the Amish example, Don Beam was having his own "engagement" with Amish culture in rural Ohio. Having fallen in love with an Amish woman, Don entreated her father, who happened to be the bishop, to allow him to marry his daughter and become a member of the church district. Don wasn't exactly a typical Amish suitor—he had attended a Catholic church as a child and had graduated from Ohio State University with a bachelor's degree in natural resources—but the congregation agreed to let him have a "proving period" of nine months. Don, who at the time was "pretty much fed up with modern-day society,"[3] subsequently became one of few outsiders to join an Amish church district, and he and his wife built a home where they would raise their four children. Working as an agricultural technician in a soils lab, Don grew very knowledgeable about plants in the remnant prairies of northeast Ohio, becoming the go-to person in the region for growing native prairie species.

Beam's encounter with Amish culture and ecological ethics, however, did not end as happily as that of the Sleeths. In fact, his passionate interest in the environment caused him, over time, to rethink his commitment to Amish beliefs. According to a biographical profile, "while the Amish way of life seemed intimately tied to the soil, the doctrine of detachment from the concerns of the world, without consideration for the future, created a conflict within Beam. His education, scientific training and natural bent had prepared him for a more active role in ecological issues."[4] After a few years, he decided to leave the Amish church, though, in a highly unusual arrangement, he continued to live with his wife and children, who remained Amish.

We spoke with Beam after he'd left the Amish—and shortly before a heart attack tragically claimed his life in 2013—and asked him to reflect further on his decision:

> The reason I got out of it was because of the lack of environmental-type awareness. Or just picking up on values that support cashing in on the environment. Cutting trees down or timbering, that sort of thing. And it's those greater issues that I found conflict with, that [the Amish] weren't knowledgeable about them, and they were using relatively close to home excuses to promote their welfare versus the greater good. And that's the part that just really bothered me. . . . And, you know, I'd try to relate things to them, like wetlands, and why we're doing this, this and this, and it's just like, wow . . . am I just supposed to remove myself from all this and ignore all these issues in the world? And essentially you are! So I just felt like I was in a cage. I had to ignore all these environmental things.[5]

Beam emphasized that his departure from the Amish church was amicable and that he was positively inclined toward many aspects of Amish life. Ultimately, though, he was never able to reconcile his own environmental sensibility with his perception that Amish beliefs and practices were insensitive to ecological concerns and even causing harm to the earth.

How could two well-intentioned parties come to such radically different conclusions about the relevance of the Amish for charting a pathway to a more sustainable and earth-friendly lifestyle? To be sure, the cases differ in important respects: the Sleeths were appropriating Amish values, as they saw them, without the benefit of closer inspection, while Beam was attempting to reconcile his own preconceptions with his lived experience in an Amish community. But we believe their stories reflect a deeper cultural ambivalence about the Amish and echo the mixed reactions that surface repeatedly in American portrayals of the Amish over the past century—the idealization of the Amish as models for how the rest of us should live, on the one hand, and outright dismissal of the Amish way of life as limiting human potential, on the other.[6] As it turns out, the contrast between the Sleeths' claim that Amish lifeways support environmental concerns and Beam's conclusion that the Amish lack environmental awareness parallels a dichotomy in media and academic portrayals of the Amish relationship to the natural world.

Our Research Questions

Images of the ecological Amish—of a wholesome, authentic people who are uniquely in tune with nature—are widespread in American society.[7] Calendars and postcards with scenes of Amish buggies or children framed against bucolic, pastoral landscapes are ubiquitous at Amish-themed tourist destinations. High-end restaurants market Amish chicken to foodies for premium prices, while Amish-grown, all-natural produce has found a niche at farmers' markets in many parts of the country. An Amish-grown strawberry is assumed to be organic and extra tasty, so much so that genuine disappointment may result if the product turns out to be only average. Hand-crafted Amish furniture is regarded as of higher quality because it is assumed to reflect sustainable forestry practices and thus to represent the antithesis of industrial production. So powerful a symbol of wholesomeness and authenticity has the Amish brand become that non-Amish businesses often appropriate it in their own advertising, relying on "the shorthand of 'Amish' to mean homespun products of yesteryear's farm family."[8]

Underlying the image of the ecological Amish is a widely held perception among the non-Amish that fundamental differences exist in how Amish and non-Amish relate to the land and to nature more broadly. Because the Amish don't spend much time interpreting themselves to outsiders, many non-Amish form their opinions on the basis of observations of Amish behavior—and they usually see a cultural group

Many non-Amish businesses use the association between the Amish and naturalness to market their products to consumers in an appealing way. Photo by Marilyn Loveless

that is rural and agricultural, eschews technology, lives at the pace of the buggy, disdains consumerism, and is centered on work and family. Understandably, they further assume that the motivations behind Amish lifestyles must be similar to those of the few non-Amish who have opted out of the dominant economy and live off the grid: they must be ecologically or environmentally sensitive or committed to sustainability.

Hence, the idea that the Amish are ecologically attentive is an assumption that springs from non-Amish interpretations of the life choices made in Amish communities. Those interpretations, however, may not accurately reflect the reasons behind Amish behaviors. The similarities with the idealization of Native Americans in the 1970s, epitomized in the "Crying Indian" commercial, which presented the actor Iron Eyes Cody as the noble, indigenous ecologist, are striking. In the Amish case, however, the dominant assumption is of a farmer who, to borrow language from Shepherd Krech III's *The Ecological Indian*, "understands the systemic consequences of his actions" and who "takes steps to conserve so that earth's harmonies are never imbalanced and resources are never in doubt."[9]

Support for the idea of the Amish as practitioners of environmentally sound land management goes well beyond the multimillion-dollar tourist markets in the larger Amish settlements. Prominent environmentalists and scholars have held up the Amish as models for a way of life that is local, self-sufficient, and in harmony with the earth. Wendell Berry, a spokesperson for small-scale agrarianism in the United States, commented that Amish farms "give the lie directly to that false god of 'agribusiness': the so-called economy of scale. The small farm is not an anachronism, is not unproductive, is not unprofitable. Among the Amish it is still thriving."[10] The writer Barry Lopez, in an otherwise scathing indictment of the environmental and cultural legacy of Columbus and the Spanish conquistadors, ends on a hopeful note: "If we are looking for some better way to farm, we need look no further than the Amish and Mennonite communities for that kind of intelligence."[11] And the author Barbara Kingsolver has celebrated the Amish farmer David Kline, who in his acclaimed book *Great Possessions* describes "a kind of farming that has been proven to preserve communities and land and is ecologically and spiritually sound."[12]

Among scholars, the late John Hostetler was a leading proponent of the view that Amish society rested on the core value of life in harmony with nature. Nearly a half century ago, Hostetler argued that "soil has for the Amish a spiritual significance."[13] In Hostetler's view, implicit in the creation story to which the Amish subscribe is the idea that the physical world is good and "the beauty in the universe is perceived in the orderliness of the seasons, the grandeur of the heavens, the intricate world of growing plants, the diversity of animals, and the forces of living and dying."[14] Hostetler's belief that the Amish would die out if they ever moved away from farming has so far proved to be inaccurate, but many other scholars have since concurred with him that the Amish provide a compelling example of sustainable agriculture that is small scale, self-sufficient, and community oriented.[15] Taken together, these accounts suggest that the broader society has much to learn from the Amish about how to enact a fundamental, philosophical shift in how we relate to the natural world.

Juxtaposed to these portrayals of the Amish as models of environmental stewardship, however, is a chorus of more critical voices. Some of the cracks in the portrait of the Amish as uniquely in tune with nature have come from local or national media coverage of Amish involvement in environmental conflicts. For example, Amish farming practices in Lancaster County, Pennsylvania, have been targeted by the Environmental Protection Agency as a major source of nutrient runoff that degrades watersheds, contributing to the "dead zone" in the Chesapeake Bay.[16] A BBC News report in 2008 focused on correcting the misperception that all Amish farmers practice organic farming by interviewing Amish growers of genetically modified crops.[17] Amish deer- and dog-breeding operations have received negative media coverage for their inhumane treatment of animals, while the threat of *E. coli* from the outhouses of conservative Amish groups has at times created conflict with county health departments. And given that the percentage of full-time farmers has fallen to under 20 percent in all the large Amish settlements, does it make sense to continue to hold up the Amish as purveyors of agrarian wisdom?

Several academic studies have bolstered the perspective of the critics. One study conducted in St. Lawrence County, New York, found that Amish farms were indeed smaller in scale, more diverse, and less integrated into the market economy than those of non-Amish neighbors who farmed, but the authors also noted high use of petroleum-based fertilizers and pesticides (including a known carcinogen, atrazine) among Amish farmers.[18] The Dutch scholar Martine Vonk has argued that direct energy use among the Amish is considerably less than among non-Amish, but she has also noted that their "rapid population growth influences the environmental impact of the Amish community significantly."[19] Surveying these and other discrepancies, the Mennonite historian Royden Loewen criticizes the tendency in writings about environmental attitudes among Amish and Old Order Mennonites to only "offer positive interpretations of these anti-consumption, technologically wary people."[20]

Perhaps the most critical frame of analysis comes from scholars who have examined the expansionist tendencies of religion-infused agricultural societies. While numerous examples exist from across the globe of ways that religious practices can harm air, land, water, and wildlife,[21] the historian Lynn White Jr. argued in a classic 1967 article in the journal *Science* that Christianity is the "most anthropocentric religion the world has seen."[22] By attacking more nature-friendly paradigms embraced by indigenous peoples, such as animism, Christianity "insisted that it is God's will that man exploit nature for his proper ends."[23] This hostility toward unhumanized nature found expression in biblical views of wilderness, as the historian Roderick Nash notes: "If paradise was early man's greatest good, wilderness, as its antipode, was his greatest evil. In one condition, the environment, garden-like, ministered to his every desire. In the other it was at best indifferent, frequently dangerous, and always beyond control."[24] While White's thesis has been much debated, a logical extension of his position would see the Amish as a prototypical example of Abrahamic, pronatalist, agricultural, settler societies whose lifestyles harm the natural environment in spite of the ethic of stewardship some claim is foundational to these traditions.[25]

The Amish Encounter with Nature

Are the Amish ecological saints or environmental sinners? We believe that posing the question as an either-or choice between these two extremes greatly distorts the complexities of the matter. We are equally convinced, however, that a rich field for inquiry exists between these two opposing interpretations. At the very least, the competing perspectives mentioned above suggest a much more complicated Amish orientation to nature and to environmentalism than is typically envisioned. They also suggest that the categories the non-Amish world imposes on Amish society may not accurately reflect the views of nature held by the Amish themselves. In this book we explore Amish lives and livelihoods in all their diversity in order to better understand their ecological imagination, their behavioral interactions with the natural world, and the relevance of the Amish for the broader effort to promote a sustainable world.

From Persecuted Farmer to Frontiersman to Entrepreneur

The Amish are often held up as a "persistent people" who follow tradition and resist change, and yet their history is more accurately characterized by adaptation to changing landscapes and livelihoods. Persecuted for their beliefs as radical Reformationists, including their rejection of the practice of infant baptism, the early Anabaptists in Europe fled from urban centers and scattered into rural hideaways.[26] Then, as the historian Leroy Beachy notes, once the Amish emerged as a distinct group in the late 1600s near present-day Berne, Switzerland, "every succeeding generation had been on the move."[27] The Amish who fled to isolated valleys in the Alsace region of France came to be known for their uncanny ability to grow crops on steep hillsides, where their livelihoods depended on nature. Yet they too were subjected to a governmental eviction decree and moved to other safe havens before several hundred ancestral families crossed the Atlantic and settled in the Blue Mountains in eastern Pennsylvania.[28]

When the first sizeable group of Amish arrived in Pennsylvania around 1730, at the invitation of the Quaker governor, William Penn, their horizons and prospects changed dramatically for the better. They joined other conservative religious groups, including Quakers, Mennonites, and those affiliated with the Lutheran and Reformed churches, who made their livelihoods off the land. Eventually, they capitalized on a frontier stretching into western Pennsylvania and then Ohio, often coming into conflict with Native American groups or settling land that the government had just opened up. In their migration west, the Amish exhibited the same frontier mentality as other white settlers at the time. But they were unlike their colonial neighbors in two key respects. For one, the contrast between Europe, where they had been hemmed in, hunted, even martyred, and a place where, released from constraints, they could expand and create their own lifeways could not have been starker. For another, the Amish came to British North America as free labor, unlike roughly three-quarters of the early immigrants. This unusual status was crucial to the economic success and growth of their communities.[29]

And grow they did. From an initial population of around five hundred settlers in the mid-eighteenth century and a subsequent wave of about three thousand from 1815 to 1860, the Amish population increased to six thousand by the early twentieth century. In 2018 the Amish number approximately 330,000 in thirty-one states and four Canadian provinces.[30] Their population-doubling time, about twenty years, is on par with that of other "high fertility" groups around the world and stands as "an expression both of religious convictions and of a people whose economy is based on agriculture and other manual trades where the labor of children is valued."[31] Over the past half century, however, the mechanization of agriculture and rising land prices in the wider society have led to an unprecedented exodus from farming. By 2010, in most large settlements the percentage of heads of household whose principal income derived from farming had declined to less than 20 percent.[32]

The ascendancy of Amish working in small businesses or factories (hereafter referred to as shop culture) has important implications for how the Amish interact with the natural world. For a variety of reasons, Amish enterprises have been very profitable, increasing overall affluence in many settlements.[33] Amish who work in factories or shops may complete their shift by midafternoon during the work week and often have weekends off. The additional free time and money created by the changing occupational structure has led to new twists on old outdoor hobbies, new avocations, and greater opportunities for travel. The rise of shop culture has also created a level of economic stratification that worries many church members, who now see extremely well-to-do Amish families living only a short distance away from those that could be eligible for food stamps.[34] Successful Amish businessmen, many of whom believe business and church matters should be kept separate, often push the envelope in redefining an appropriate relationship to nature, whether through high-end horse breeding, trips out West to hunt big game, or extended vacations in the Pinecraft, Florida, settlement.

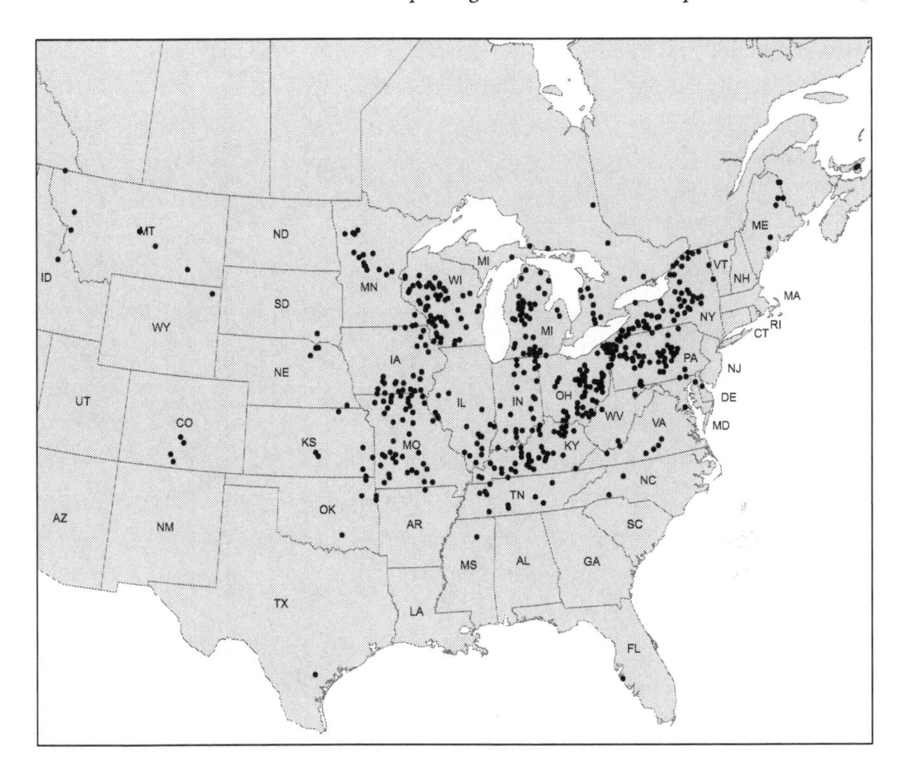

Amish Settlements in the United States, 2018. Map by Ric Reynolds

Biblical Literalism

The Amish are a Christian Anabaptist (literally, "rebaptized") group because they insist that baptism takes place after one reaches the "age of accountability" (usually, sixteen years old) and thus reflects a conscious commitment to follow the example of Jesus Christ and the biblical word of God. Amish spirituality is grounded in several authoritative texts, including the *Martyr's Mirror* (which chronicles the persecution of eight hundred individuals killed for their beliefs in the 1500s), the *Ausbund* (the oldest Protestant hymnal still used today), and several prayer books. The Bible, however, remains their cornerstone, and the Amish "have an abiding confidence in the Bible as an entirely authoritative guide for daily living."[35] Unlike those Christian groups that interpret the Bible through historical and critical analysis, the Amish believe the Bible should be taken at face value as the word of God and, when unclear, interpreted primarily by ordained leaders or by the entire congregation. This belief in biblical literalism had its origins in the Protestant Reformation, when reformers needed to appeal to a higher authority in their efforts to oppose the Catholic hierarchy, but it also has important environmental implications.

This biblical literalism shapes Amish views of humans' relations with nature in fundamental ways. In the creation story, for example, God sets humans apart from nature and gives them control over the resources of the earth. The rainbow in the sky indicates "God's covenant" to never again destroy the earth by flood and stands as a reminder of both sin and the possibility of redemption.[36] The book of Revelation predicts that after heavenly signs of a world in crisis, marked by famines, disease, and war, control of the world will be gained by an antichrist, backed by a false prophet who carries the number 666, the mark of the devil. We ask how the Amish think about the environmental implications of these biblical texts. Do they debate whether Genesis implies dominion or stewardship over earth's creatures? As "pilgrims and strangers in this world,"[37] do they direct their choices primarily toward the world to come instead of toward the care of this earth? Even within a foundation of biblical literalism, Amish individuals and communities may weight biblical texts in different ways.

Another implication of biblical literalism is the Amish tendency to make "pre-scientific assumptions about the natural world."[38] Ironically, the scientific revolution had its origins in western Europe in the sixteenth and seventeenth centuries, at approximately the same time that the early Anabaptists were being persecuted for challenging the corrupt practices of the Catholic Church. In the course of their historical odyssey, however, the Amish have parted ways with certain aspects of modern science, to the point where the subject is avoided in their parochial schools and "evolution is considered heresy."[39] Yet most Amish rely on the fruits of science on their farms and in their businesses, recreational pursuits, and health care. In addition, they live in a society in which science has been central in shaping the modern understanding of how nature works, as well as "what counts as environmental problems."[40] We ask how science becomes appropriated by the Amish in their everyday lives and how they maintain their view of a "young earth" and of the Bible as the authoritative and literal word of God even as they increasingly rely on technologies developed through evidence-based science. We also explore the implications of the limited role of science in their parochial schools for their use of natural medicines and for their awareness of environmental problems.

Separation from the World, or Deliberate Marginality

Few aspects of the Amish world-view are better known than their attempt to distance themselves from worldly matters, as instructed in biblical texts such as Romans 12:2,

"Be not conformed to this world," or 1 John 2:15, "Love not the world, neither the things that are in the world." When Amish youth kneel in front of the congregation to be baptized, they renounce the self, the devil, and the world. Though the Pennsylvania Dutch word for the Creation (*Schopfung*) has a decidedly positive connotation for the Amish, the term for "world" (*veld*) invokes images of the vices of mass culture and sinful excesses. At the broadest level most Amish can agree that *Absonderung*, or "nonconformity" to the world, "means keeping a cautious distance from evil: violence, war, promiscuous sex, abortion, greed, fraud, divorce, drugs . . ." as they "seek to follow Jesus on the narrow path."[41] The reality, however, is that Amish leaders frequently disagree about how to define separation from the world, resulting in a wide diversity of accommodations to modernity among Amish groups today.[42]

The ecological implications of the doctrine of separation from the world are far reaching, and they include attitudes toward technology. Despite claims that the Amish are antitechnology, they in fact "selectively sort out what might help or harm them."[43] Their responses to technology include rejection, acceptance, adaptation, invention, and distinctions between ownership and access. In addition, all Amish groups observe a "spectrum of restraint," with technology most restricted in schools and homes and on farms and somewhat less so in shops or outside employment. In most cases, the overriding concern is how technology will affect the quality of family and community life, a balancing act that has become increasingly difficult because the world confronted by Amish families today is a far cry from what it was even fifty years ago. The result of these restrictions on technology, especially forbidding ownership of automobiles and electricity from the grid, is a lifestyle that limits access to some technologies that have major carbon costs. Precisely how the Amish ecological footprint compares with that of non-Amish living in a similar rural setting remains an open question, one that we explore in chapter 3.

The legacy of nonconformity extends to Amish views of government, which were powerfully shaped in sixteenth-century Europe by the state's response to the early Anabaptist critics of corrupt practices in the Catholic Church. Because of their belief that "they should live a disciplined life accountable to one another rather than to the state church," approximately twenty-five hundred Anabaptists were hunted down and killed in gruesome ways between 1527 and 1614.[44] Their history of persecution by the state in part explains why contemporary Amish will not serve on juries or in the military, run for political office, or file lawsuits, preferring instead a minimalist state that restricts itself to maintaining civic order. While the Amish are especially uncomfortable with the twentieth-century expansion of the "warfare state" and the "welfare state," they are also deeply suspicious of the increased governmental regulation of land, water, air, and other natural resources. As the American environmental movement has taken on an increasingly secular and scientific bent, regulatory pronouncements have become the tool of choice for extracting compliance from a diverse citizenry. In this context, we ask how the Amish imagine the government's motives and react to governmental efforts to protect the environment.

A Guiding Framework

Broadly speaking, our study rests within the interdisciplinary theoretical framework of political ecology, which examines the natural world as the site of a confluence of powerful political, economic, and sociocultural forces.[45] We draw on four central concepts in the field that we find particularly helpful in illuminating the Amish case.[46] First, we see nature as invested with cultural meanings and symbols, and therefore we explore Amish constructions of nature as they occur in a wide variety of social contexts. We complement this symbolic approach with attention to structures of power and inequality, asking whose interests are served and whose are compromised by the changing economic livelihoods of Amish individuals and their appropriation of natural resources. Third, we emphasize the importance of historical context by highlighting the multiple ways Amish are positioned in relation to the natural world and how these positionings have changed over time. Finally, we examine the natural environment and the biological organisms that surround Amish settlements, asking how available ecological resources shape Amish lives and how Amish choices affect those same ecosystems.

Culture

The moral values and interpretive frameworks offered by religion and culture are crucial determinants of the way societies and people imagine their environment and the challenges of sustainability. Rather than relying only on outsider categories and explanations, we seek to understand the webs of significance the Amish themselves create around nature and the knowledge they have about landscapes near and far. We draw on insights from the field of ethnoecology, which, according to its founder, Harold Conklin, attempts to understand how "environmental components and their interrelations are categorized and interpreted locally."[47] Because so little has been written about Amish views of nature, one of our main goals has been simply to listen, observe, and record—to discover how the Amish talk about nature, how it forms the basis of their livelihoods, how it figures in their recreational activities and finds expression in their writing and other creative outlets. We began with

Amish engagements with the natural world are gendered in many ways but also present examples that defy outsider expectations, such as this young woman baling hay. Photo by Doyle Yoder

fundamental questions. Do the Amish see themselves as part of nature or distinct from it? Do the Amish moralize nature, and do they even consider themselves to be "ecological"? How exactly do environmental concerns enter into the Ordnung of church districts?

A focus on cultural meaning leads us to examine the representations and uses of nature in Amish life and to explore how Amish activities draw on and impact their natural surroundings. We consider the significance the Amish attach to particular animals and plants, from horses, snakes, and purple martins to burdock leaves and morel mushrooms—as well as the ways they read signs of nature to intuit divine will. We also examine how cultural assumptions about gender, age, or status influence Amish interactions with nature. The Amish believe that men and women, though equal in the eyes of God, have separate callings, yet virtually no research exists on how the gendered dimensions of their lives map onto their use of natural resources for livelihood, recreation, and health.[48] What kinds of interactions with nature are seen as appropriate for boys and girls, men and women? An ethnoecological approach demonstrates the importance of understanding how the Amish themselves symbolically construct nature and how their distinctive values and version of Christianity shape their interactions with the physical and biological environment.

Political Economy

A political-ecology framework argues that cultural world-views are always embedded in a landscape of power relations, including structures of inequality that may change over time. This perspective reminds us that we should not study the Amish in isolation but rather attend to how their choices about natural-resource use are perceived by and impact other groups. It also prompts us to ask how attitudes in the broader society have shaped Amish views of nature, since, as Steve Nolt reminds us, "today's Amish world is in part a product of identities crafted largely by others."[49] Perhaps because the Amish themselves often avoid engaging in certain aspects of the political process, such as voting, serving on a jury, or joining civic organizations, most studies of the Amish tend to shy away from an overtly political frame of analysis.[50] Yet the choices the Amish make about land purchases and land use, as well as the way they organize their farms, businesses, and outdoor recreational activities, constitute a landscape of practice that has social, political, and ecological implications. Through numerous committees that interface with local, state, and federal agencies, the Amish do in fact engage in informal lobbying on policies related to agriculture, hunting, animal husbandry, forestry, zoning, and much more. We examine what kinds of policies they support and what discourses around nature, society, and the environment they promote.

The Amish are a distinctly rural people, but rural-urban distinctions also exist in a field of power relations. As Barbara Ching and Gerald Creed have shown, in the United States a cultural hierarchy operates: "rustic identities" (those associated with the countryside) are marginalized and denigrated, while "urbane or sophisticated identities" (those associated with the city) are celebrated.[51] In their view, the common oppositions that shape views of the urban-rural divide—educated versus uneducated, secular versus religious, middle or upper class versus working class—have had a profoundly negative impact on rural people, leading them to form counternarratives to the perception of urban elitism. Yet while the Amish do "hold a strong bias against city life,"[52] they complicate this cultural hierarchy. As a rural people who often put down roots and pass land on to their children, the Amish may come to have a deep familiarity with their immediate locale, a valuing of place that non-Amish environmentalists have lauded.[53] At the same time, ecological processes are not limited by property lines or by community boundaries. Some of the most pressing ecological issues facing the world, such as the supply of clean water, require an ability to think and act beyond the borders of any one community. To the extent that the Amish are uninterested in or unwilling to move beyond local knowledge and the interests of their own communities, can we really say that they are ecologically minded?

Every society produces mental conceptions of what constitutes the appropriate use of space and organizes its particular landscape in a tangible way.[54] By extension, as the French philosopher and sociologist Henri Lefebvre argued, when one society comes into contact with another, the spaces it produces "interpenetrate, shove aside, or shatter the spaces of people already living there."[55] Most studies using Lefebvre's concept of the "production of space" have focused on how capitalism and its logic of accumulation and consumption have uprooted, transformed, and destroyed indigenous cultures that made their living off the land. In applying this frame to the Amish, we ask how the increase in population density in the large Amish settlements and the expansion of the Amish into new regions alters the landscape in ways that are both visible and invisible. If, as leading scholars of Amish life maintain, it is possible to identify an "Amish way," a constellation of core religious and cultural values, then how are these guiding principles enacted spatially, and with what ecological effects?[56]

History

A political-ecology framework also emphasizes how human interactions with the natural world change over time. Our interest lies in how the most recent historical shift in economic livelihoods has reconfigured the Amish relationship to nature. For the minority who have remained farmers, we ask how they have navigated the organic and farm-to-table movement, the rise of genetically modified crops, and new governmental regulations for food safety and soil health. And for those who have embraced shop culture, either running their own business or taking a factory job, to what extent do their livelihoods and leisure activities still depend on their ability to read the landscape, to adapt to natural events, and to extract natural resources? While Amish children don't grow up with the same set of expectations and anxieties around high school, advanced degrees, and professional career paths that characterize college-bound, non-Amish youth, their economic fortunes are increasingly tied to regional, national, and global markets. Recognizing that globalization has created a complex mix of affluence and poverty in rural America since the 1980s,[57] we ask how the Amish are relating to nature in new ways, as animal breeders, herbalists, outdoor recreationalists, travelers, and writers.

Another historical shift with implications for understanding the Amish relationship to nature is the rise of the environmental regulatory state in the early 1970s, the culmination of a heated public debate between conservationists and preservationists that started in the early 1900s. Led by President Theodore Roosevelt and the head of the United States Forest Service, Gifford Pinchot, conservationists argued that federally owned land and national parks should be managed efficiently for human use and recreation. "The first great fact about conservation is that it stands for development," argued Pinchot in his 1910 book, *The Fight for Conservation*.[58] Eloquently articulating a preservationist stance was the naturalist John Muir, who argued that humans were intruders in sacred natural spaces and that timbering, grazing, and dam building should be off limits in designated wilderness areas. Muir's ecocentric approach can be seen in his attack on proponents of the Hetch Hetchy Dam in Yosemite National Park, whom he described as "temple destroyers, devotees of ravaging commercialism [who] seem to have a perfect contempt for Nature and, instead of lifting their eyes to the God of the mountains, lift them to the Almighty Dollar."[59] Muir's Christian overtones and scriptural style make him sound like someone who would appeal to Amish sensibilities, but as we demonstrate, contemporary Amish views of nature share more similarities with those of early-twentieth-century conservationists than with those of the preservationist school of thought.[60]

A final historical shift relevant to our study is the landscape of religious fracture that has arisen among the Amish since the early 1900s.[61] The number of affiliations, or "loose federation[s] of like-minded Amish churches," now stands at more than forty across North America.[62] Their names, such as Swartzentruber, Lancaster, or Old Order, come from a confusing mix of surnames, geographic identifiers, and labels applied to them by others.[63] Many Amish express regret at these schisms, which the historian Leroy Beachy describes as a "seemingly reckless century of division."[64] Each affiliation has carved out a set of identifiers to justify its separation that includes technology use, home architecture, occupational types, and dress and transportation styles.[65] With some caveats, the different Amish affiliations can be placed along a spectrum in terms of their degree of worldliness. The Amish themselves use the value-laden terms "low" and "high" to compare conservative and liberal affiliations. In general, lower affiliations have more limits on technology use and observe stricter shunning and separation from the world, while more liberal affiliations make more compromises with technology and "emphasize a more personal and reflective religious experience."[66] Because these affiliations function like competing status groups, Amish identity today is as much a product of one's relation with other Amish groups as of one's relation with "English society."

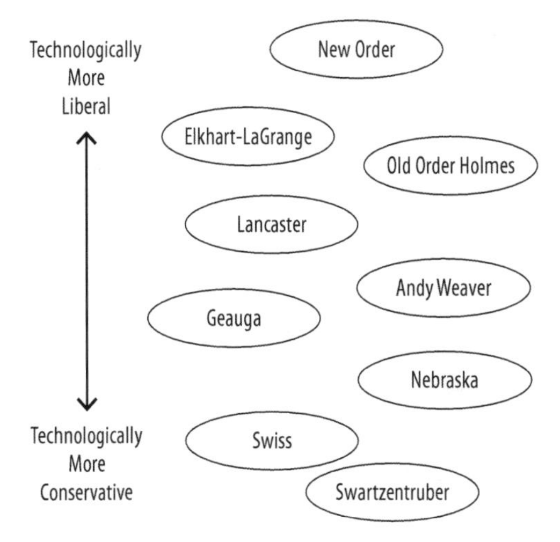

Figure 1.1. A pictorial representation of affiliations mentioned in this book based on their degree of accommodation to technologies and other worldly influences. By Marilyn Loveless

The specifics of affiliations can seem confusing to outsiders, but the distinctions are important because the Ordnung of a church district shapes the tools at its disposal for harvesting and marketing natural resources, from the kinds of energy and travel allowed to the possibilities for outdoor recreation and land use. In the following pages, we frequently reference a handful of Amish affiliations. The New Order Amish are among the more liberal in accepting technology, but they are theologically conservative on issues of courtship and rumspringa. The Elkhart-LaGrange Amish (mostly in northern Indiana) have a very liberal overall reputation; they are followed closely by the Old Order Amish of Holmes County, Ohio. Lower on the spectrum, the Lancaster Amish of Pennsylvania and neighboring states, as well as the Andy Weaver Amish and the Geauga Amish of northeastern Ohio, are moderately conservative on shunning and other theological issues but fairly liberal on accommodations for businesses. The lowest, or most conservative, groups, those that have tried to resist the encroachments of modernity in many respects, are the Swartzentruber Amish (originally in Ohio but now in fifteen states), the Nebraska Amish in Big Valley, Pennsylvania, and the Swiss Amish in Indiana (see figure 1.1).[67]

Nature

Finally, a political-ecology framework takes seriously the biophysical realities that constitute a given landscape or region. Beginning with Bill McKibben's *The End of*

Nature in 1989, followed by William Cronon's 1995 essay "The Trouble with Wilderness," a number of prominent environmentalists and scholars have argued that nowhere on earth has the natural world remained untouched by human activities and therefore nature itself no longer exists apart from human intentions and actions.[68] As mentioned earlier, humankind has been judged by some to have crossed the threshold into a new epoch, the Anthropocene,[69] in which our collective behavior influences, though does not yet completely control, all aspects of the physical environment, including wilderness and weather.[70] The notion "humanized nature" reflects this expansion of the human footprint and refers to "a nature that is a by-product of human conceptualizations, activities, and regulations."[71] Moreover, Cronon is partly right that we need "to abandon the dualism that sees the tree in the garden as artificial—completely fallen and unnatural—and the tree in the wilderness as natural—completely pristine and wild."[72]

Yet the extreme constructivist argument that nature itself no longer exists apart from human activities is also problematic.[73] In spite of the pervasive influence of humans on the natural world, places still exist where direct human disturbance is limited. Organisms still have their own life histories, and ecosystems their own dynamics; forests, for instance, have an "unmetaphorized reality" that must be recognized.[74] Laws of chemistry and physics have not been invalidated. In other words, there is a finite and corporal universe that would exist with or without humans. The ecological anthropologist Roy Rappaport aptly captures this notion of nature's capacity to act independently of our understanding of it: "Nature is seen by humans through a screen of beliefs, knowledge, and purposes, and it is in terms of these images of nature, rather than of the actual structure of nature, that they act. Yet it is upon nature itself that they do act, and it is nature itself that acts upon them, nurturing or destroying them."[75] For those who accept evolution, the idea of nature as an independent entity is premised on its long existence before humans arose. Although human activities today may be altering selective pressures on organisms, nature will exist after humans are no longer on the planet. But if nature's independent existence is dictated by theology, as it is for the Amish, then such thinking will have ramifications for environmentalism.

Taking seriously the dynamics of ecosystem processes also means relying on science, which describes ecological phenomena on spatial and temporal scales that humans sometimes find difficult to grasp. Landscapes vary in the resources they provide and thus make possible or constrain the ways that nature can be appropriated and used. But ignorance of science or suspicion about scientific statements may color the ability of Amish to make ecologically sound decisions. We explore

Amish households in the western United States, such as this one in Rexford, Montana, face a different set of ecological opportunities and constraints than settlements back East do. Photo by Marilyn Loveless

Amish views about their duty toward nonhuman life forms and Amish knowledge of ecosystem dynamics. And in considering the varied ecological contexts in which Amish settlements exist, we ask what elements of their interactions with nature are common across their communities and how they adapt to the resources and opportunities unique to particular landscapes.

The Amish are exposed to the natural environment to a degree that is uncommon for most other twenty-first-century Americans. Although Amish homesteads are in human-altered landscapes, they still come with a full quota of plants, animals, and soil, starry skies and dramatic cloudscapes. Buggies travel at a relatively slow speed, with a soundtrack of horse hooves, bird calls, and the trill of insects. While most of us pass through our surroundings at dizzying speed, sealed inside an automobile listening to the radio, the circumstances of Amish lives give them access to their nonhuman environment at every turn. Many of the rest of us have a suspicion, at least now and then, that we are missing something by living much of our life indoors. So it is intriguing to examine how the Amish capitalize on this proximity to nature—how they interact emotionally, intellectually, and behaviorally with their natural surroundings.

At the same time, no human population has a natural propensity for living in harmony with its environment. No group, including the Amish, possess an innate ecological wisdom, as if this capacity were somehow hard-wired in their genes. Although members of small-scale foraging and agricultural societies often do develop an intimate understanding of the flora and fauna that surround them, even that collective ecological knowledge can be lost in just a few generations during periods of rapid cultural change. As white settlers of German and Swiss heritage who arrived in colonial America in the 1700s and 1800s, the Amish can hardly be equated with the indigenous peoples of the Americas, and yet their relationship with nature is often idealized in similar ways. But identifying "ecological ethnicities"[76] is not "as easy as pointing to non-industrial peoples."[77] We need to take a more analytical approach. In the chapters that follow, we try to understand in some detail how nature is negotiated by the Amish as a lived landscape. In doing so, we hope to critically reflect on the elements of belief and practice that contribute to living lightly on the earth.

I *Growing Up Rural*

 Raising Children at Nature's Doorstep

In the summer of 2014 the Canadian subsidiary of Nature Valley granola bars launched an advertisement entitled "3 Generations" that featured grandparents, parents, and children from the same families responding to one question: What did you like to do for fun as a kid? The two older generations shared memories of time spent in the outdoors—berry picking, sledding, fishing, and building forts. When it was the children's turn, however, the answers, as well as the music and the mood, changed dramatically. One girl related how she watched an entire TV series in just four days. Another confessed, "I would die if I didn't have my tablet." The contrast between the way today's "indoor children" spend their time and the outdoors-oriented childhoods of their adult mentors could not have been sharper. Perhaps unsurprisingly, some reactions to the ad campaign were quite critical. Online news outlets took Nature Valley to task for shaming parents, vilifying technology, and implying that the solution to the problem lay simply in buying more granola bars. Nevertheless, the themes depicted in the ad clearly hit a chord; it received more than half a million hits in under two weeks.[1]

The "3 Generations" ad highlighted a well-documented cultural transformation: the time American children spend outdoors is low and declining.[2] Though some children still participate in scouting, attend nature camps, and play freely outdoors, the late historian Thomas Berry argued that as a society, we have become autistic in our encounters with the natural world.[3] The journalist Richard Louv, in *Last Child in the Woods*, coined the term "nature deficit disorder" to describe the implications of the estrangement from the natural world that characterizes American children today. While one can debate the merits of a technology-mediated childhood spent largely indoors, the growing level of inattention to natural cycles and loss of curiosity about the natural world does not bode well for the future of conservation.

This chapter explores the degree to which Amish children experience nature differently from many non-Amish children in the twenty-first century. Does the general portrait of a denatured childhood hold true for Amish children as they grow up at home and complete eight years of formal schooling? Or do they embody the

The rural locations of Amish church districts and the absence of homework in Amish schools afford ample opportunities for children to play outdoors once chores are completed. Photo by Doyle Yoder

ideal that the Amish author David Kline expresses when he writes, "All children should have a creek running through their childhood"?[4] The geography of an Amish childhood creates many opportunities for outdoor activities. In the process of doing chores, playing outside, and walking to school, Amish children directly experience the weather, plants and animals, and other natural phenomena. Yet within the walls of their Amish parochial school, nature study takes on a peripheral role, and science is largely absent from the curriculum. Thus, Amish children enter adolescence with an intimate knowledge of many aspects of the landscape in which they live but minimal awareness of how science works and what it can tell them about the world around them.

The Horizons of Amish Childhood

The spatial ecology of Amish life actively promotes outdoor activities and shapes children's views of nature in important ways. First, Amish homes are not typically hemmed in on small, quarter- or half-acre plots as many non-Amish homes are. Most have big yards, and children can roam widely on their own or their neighbors' land. Even nonfarming Amish families want at least a two-acre plot with a barn and a small pasture for the horses, and most homes have additional outbuildings, so daily activities require going outside. Changes in the weather are a felt reality, especially for the conservative affiliations, who heat their Spartan homes with wood. Recalling snow that sifted under the window sash and piled up on the floor the previous winter, a Swartzentruber man quipped, "But sleeping in the cold is healthy." Homes generally lack air conditioning and electric lighting, so the Amish are accustomed to heat, humidity, and seasonal changes in daylight. Because their rural communities are less impacted by light pollution, the Amish are very aware of the night sky. The *Gemeinde Register*, a church newsletter, carries a monthly "Licht Calendar," which gives the dates of phases of the moon and the positions of stars and always ends with the friendly exhortation "Keep looking up!"

Amish children's lives are spent largely free of structured recreational activities and beyond the reach of mass media. Owning a television is off-limits to all Amish groups, and internet-enabled laptops and smart phones in the home are largely restricted. Sending a child to day care or preschool is virtually unheard of in Amish communities, and Amish children seldom participate in 4-H Clubs, Little League, or programs sponsored by local recreation departments. Time that might otherwise be spent in formal organizations or in supervised recreation is thus available for unstructured outdoor play or for family gatherings and camping trips. Most Amish children attend school for only eight years, and homework is strongly discouraged. Rather than being driven or escorted by their parents, Amish children usually walk or bike to school on back roads with their peers, which provides ready opportunities to explore along the way.

For a variety of reasons, parental nervousness about "stranger danger" is minimal. "Well, I think that's the best place to put them, in the woods," noted one young Andy Weaver woman. "It's better than letting them loose in a shopping mall. You never hear of people being shot in the woods, but you do hear of that happening in the city. It's the city that is dangerous." In contrast to the hypervigilance seen among some non-Amish parents, it's not uncommon for Amish houses to be unlocked and for children to come and go as they please. One mother of seven commented, "Well, we have about three acres, and our property goes to the creek. They pretty much have free reign, they can be in our yard, or in the creek. And there's a swamp behind us, and they like to go back there. The only restriction is, they should respect people's property and not bother anything." Nor do Amish parents try to protect children from most of the unpleasant aspects of the outdoors. "In general we won't teach them about poison ivy or worry about mosquitoes, or this or that," commented a mother of eight. Direct experience is considered to be far and away the best teacher.

Moreover, mothers typically do not pursue paid employment outside the home, though they may run a small business, such as selling baked goods or flowers, from their homes. In farm families, the father is at home, and even fathers in nonfarming families will sometimes build a shop close to the home to maximize family time. Thus, Amish families largely avoid the situation of two working parents with little time to spend with their children outdoors. Children usually have multiple siblings, cousins, aunts, uncles, and grandparents living nearby, providing playmates and mentors to assist in outings or projects. The basic parameters of religious and family life thus ensure that Amish children spend a considerable amount of time outdoors.

Variations in family income and church guidelines do, however, influence acceptable forms of transportation and thus lead to different outdoor opportunities for Amish children. In some cases, families may purchase a second property for hunting or camping. We met a few successful businessmen who had rented an RV and hired a driver to take their entire family on a trip out West. Those with more modest means may visit state parks or public recreation areas. Children in church districts that allow bicycles can explore areas outside the confines of their immediate neighborhood, while those in districts that only allow scooters are somewhat more confined. One man from the Geauga settlement in northeastern Ohio had not seen Lake Erie until he was eighteen even though he had grown up only twenty-five miles from the lakeshore. But even children in the ultraconservative groups, limited to walking or riding in the buggy with their parents or an older sibling, still find time to explore their own and their neighbors' farms. "I used to collect arrowheads, I had hundreds of them," remembered an ex-Swartzentruber man. "That really intrigued my mind. . . . First white man to touch this. Or after it killed a white man!" he laughed.

How and Where Do Amish Children Encounter Nature?

The restrictions on owning televisions and home electronics open up a rich space for Amish children to learn about nature through outdoor activities. Most children are assigned regular chores, which often include taking care of animals or working

For farm and nonfarm families alike, chores often involve taking care of animals, which provides many lessons about nature. Photo by Doyle Yoder

outside. Chores bring together two of the primary habits Amish parents try to teach their children: hard work and obedience.[5] Especially in farming families, chores not only teach a strong work ethic but allow children "to see God's natural design," according to an Old Order father of six. "You see baby animals being born and grow up and produce food and how they need to be taken care of and you know where that food comes from and what the weather has to do with it. There's a lot of education that comes with it." Another father noted that even nonfarming families have horses, and "you can learn a lot about nonverbal communication from being around horses." And in the case of both horses and children, "you have to earn their respect. You have to discipline them, but they have to trust you."

Chores are often tied closely to household income-generating activities, and parents sometimes use monetary incentives to motivate children. "If you're farming, you have to get [the children] enthused so they want to help. If we sell a calf, we give them a small amount [of the profit when the calf is sold]," commented one Old Order mother. Another said she went in "half and half" with her son on raising strawberries. From a very early age children learn that nature is a resource that can be exploited for monetary gain. Chores are sometimes gendered, with boys doing more outside work, feeding the animals or shoveling manure, and girls helping with housework, such as cleaning and doing laundry. But some activities, like yard care, are shared by both boys and girls. The division of labor usually depends on the children's ages and birth order, the number of boys and girls in a family, and parental inclinations. Some activities involve working closely with parents or grandparents, providing opportunities for them to pass on knowledge. "How do we all know ash and elm are the best firewood? Because boys cut wood with their fathers," said one man. And though sex education is rarely explicitly taught, one horse breeder recounted how his four-year-old daughter yelled "Way to go, Herbie!" when his stallion successfully mounted a mare. The activities, adventures, and mishaps of "doing chores" are sources of shared experiences for Amish families, creating strong memories that are passed from generation to generation.

Amish children also engage in outdoor hobbies that combine recreation, competition, and income or food procurement. In most settlements deer hunting and fishing, and to a lesser extent birding and trapping, attract the interest of boys, while girls pursue gardening or horseback riding. Amish boys often have been given a BB gun by the time they enter school at age six, and they practice shooting sparrows and other birds, though usually "not cardinals" because "they're pretty." Many boys and girls go out hunting with their fathers as early as age seven or eight. This is especially true in the western settlements, where entire families, sometimes with relatives from back East in tow, take pack horses into the high country for one- or two-week camping or fishing trips. Older girls are less likely to hunt, but one girl in her late teens said she hunted just to show that girls could do it: "I like to rub it in that I'm the only one who got a buck last year. My brother just got does."

Apart from chores and hobbies, Amish children spend ample time in unstructured outdoor play. Climbing trees, riding bikes, building forts, and "playing Indian" are popular activities, and many involve a competitive slant. One man from the Ashland, Ohio, settlement remembered filling shoeboxes with his collection of birds' eggs, competing with friends to find the most species. A Geauga woman declared, "We would compete to see who had the hardest soles," referring to the common Amish practice of going barefoot. One woman remembered spending days digging a tunnel under a fence. "So I guess Amish kids aren't afraid of dirt," she concluded. And a woman in her twenties noted that "the neighbor boys would say girls had no business tromping in the woods and forest, but my mom understood, and she let me run wild. My neighbors were back home baking cakes and raking leaves." Asked what she played as a child, another woman said, "We played in forts, we had playhouses and teepees, we were taken prisoner, and we did skull and sun dances," apparently arranging animal skulls they'd found and dancing around them.

What Do Amish Children Learn about Nature?

As a result of frequent, direct outdoor experiences and intentional socialization by family and church members, Amish children acquire a set of widely shared understandings about their relation to the natural world. From early in life, they learn that nature reflects God's handiwork. "Yeah, God created it, you pick that up as a very small child. We did. At a very young age we're introduced to the Bible and how God creates beautiful things," commented a man from the Lodi, Ohio, settlement. One implication of God's beautiful creation is that humans should take care of the earth: "You know, Psalms 24 says the earth is the Lord's and the fullness thereof, and so that's one reason we wouldn't misuse it, because in doing so you would be working against God's creation. He made things beautiful, wonderful, it functions well, and so if you violate that, you're actually violating God's order." The ordering of nature, then, is not by accident; rather, the "laws of nature" are seen as a testimony to God's infinite wisdom. A second implication of this way of thinking, according to a woman from the Geauga settlement, is that "we can be responsible for ourselves, but we're never going to have an impact on the earth. The idea that you can make the world a better place is dangerous if you think you are stepping in and playing God." In other words, people need to know their place in the universe and not cross the boundaries established by God.

One lesson all children learn about God's intentions is clear: nature is for the benefit of humans. One New Order man put it this way: "It wouldn't be a popular thing to say in society, but Man is dominant on the earth. There's no way around that, because Man is the only part of creation that has a soul and has reasoning powers. Some birds and creatures have certain abilities, but there's really no comparison to the mind that Man has." An Andy Weaver young man expressed this perspective very directly. He told us, "I can't *ever* remember being taught by my parents about the ecological value of nature." As an example, he cited the trees on his property. He takes good care of them because they are useful to him as firewood and possibly to sell for timber, not because taking care of them is ecologically valuable. Any Amish behavior that seems to outsiders to be environmentally motivated, he said, is really just a by-product of religious values that emphasize thriftiness, simplicity, and responsible use of God-given resources. This perspective differs sharply from the assumption many outsiders make that Amish simplicity stems from a desire to protect the earth.

At the same time, nature reflects the events that occurred in the Garden of Eden, when the serpent tempted Adam and Eve to eat the forbidden fruit. The Amish tend to divide the flora and fauna into those that are beneficial to humans and those that are not. And just as in the non-Amish world, assigning evil or good to an organism often bears little relationship to its place in the ecosystem. Perhaps the most striking example of this literal reading of nature as reflecting biblical events is the Amish view of snakes. "Well, the snake is the form that the devil came in," noted one Old Order woman. An Old Order man elaborated, "Oh, yes, the Amish are scared to death of snakes because they take the bible story to heart. Eve was tricked by a snake . . . and it says if you see a snake, beat it! Amish women will go screaming if they even see a garden snake. And you don't just kill it, you smash it to smithereens." Asked what she does when she sees a snake, one mother said, "The boys take care of it," at which point her twelve-year-old son chimed in, "Take the hoe." Unsurprisingly, folklore surrounding encounters with snakes abounds. If you throw a snake into the fire while its tail is still convulsing, we were told, it will scream at you or grow legs.

Not all Amish take a "kill on sight" approach to snakes, however. Small children are considered innocent and able to commune easily with nature, as exemplified in the story of a little girl sitting on the back step of her home who innocently shared her "baby soup" (milk and bread) with a black snake. Many Amish are aware of the ecological value of snakes. "Some parents have taught [children] snakes are evil, but I'm not comfortable with that because I think snakes have a role in nature," commented an Old Order woman. It is extremely difficult to go against such a powerful cultural code, though. "Well, I know it is [ecologically beneficial], but I kill them, I do," confessed one man. "Last summer there was a big five-foot black snake in our yard, and I was the unlucky person to be at home to take care of it. So I did." One Old Order man with a strong belief in the ecological value of snakes said that other Amish in his community will actually view him negatively if he doesn't kill a snake.

Because children frequently experience nature in its raw form and grow up hearing that other creatures do not have souls, their attitude toward animals is relatively unsentimental. "Aww, you should've got him," yelled an Amish acquaintance from the back seat of our car one night as we narrowly missed hitting a mink crossing the road. Mink, he told us, are a real threat to their chickens and very hard to kill. One former Amish man who grew up in a conservative affiliation reflected on his relationship to animals then and now: "Horses were to plow the fields. They were not pets. [A buggy horse] was just transportation for me, that's how I viewed him. Cows, hogs, they were farm animals. But now we come to dogs and cats, and this is kind of bad . . . but it's what we did. Sometimes there were so many of them that we would get the gun and shoot 'em. . . . Now they live with us, and we cry when they die. I mean out here [in English society] they're family members." Still, the Amish are far from devoid of feelings toward animals. Horses are often given laudatory names, like Big John, while animals that will end up on the dinner table may have names that reflect their eventual fate, such as Bacon or T-Bone. And attitudes have changed somewhat over the past couple of generations, especially toward horses. Many children now grow up caring for ponies, driving pony carts, and training horses for recreational riding.

The Changing Contours of Amish Childhood

The reality of more and more Amish adults working in shops or factories, which has led to new levels of wealth and leisure time, has paradoxically created a desire among some Amish parents to steer their children into nature-related pursuits. "We definitely would promote [outdoor activities]," commented one New Order man. "We'd much rather have that than some of these others, especially professional sports and electronics." Also competing for the attention of Amish children in the higher affiliations are Game Boys, iPods, and other gadgets. "Battery-powered devices have been discouraged, but some families use them a lot," noted one Old Order mother. After children turn sixteen, but before they join the church, many get smartphones, which allows them, in essence, to carry the world in their pocket. As one Old Order father put it, "The countryside is not as quiet as it was thirty years ago."

In spite of these changes, Amish children have thus far largely escaped the more negative effects of a sedentary, media-saturated lifestyle, including childhood obesity and asthma. "If they had all the electronic gadgets and games to compete for that interest, it definitely would be different," reflected one New Order father. "I guess the fact that we don't have TV and all that stuff fosters an interest in the outdoors, definitely." According to the environmentalist David Orr, the Amish are an exception to one of the central trends of our time. They maintain routine, daily interaction with animals and direct contact with nature at a time when contact with manufactured things is increasingly the norm and nature has largely become an abstraction.[6] At the same time, their close contact with nature occurs within a very specific cultural idiom that sees nature in unsentimental terms, as designed by God for the benefit of humans.

With increasing affluence, swing sets, trampolines, and landscaped play areas have become more popular among some Amish groups. Photo by Doyle Yoder

The Role of Nature Study and Science in Amish Schools

When Amish children reach the age of six, they become "scholars" and begin eight years of formal schooling, concluding at age fourteen as allowed under the landmark 1972 Supreme Court decision *Wisconsin v. Yoder*. The large majority attend a private (Amish parochial) school administered by a three-person school committee made up of fathers who are not ordained leaders.[7] The purpose of the parochial-school curriculum is, first, to reinforce values taught at home and in church that are conducive to the Amish way of life and, second, to prepare children to make a living in a business environment that will require basic literacy and numeracy and interaction with non-Amish. Thus the curriculum, taught entirely in English except the High German classes, is bare bones. In all schools, the main focus is on reading, writing, and arithmetic, with history, geography, and health sometimes taught in schools that cater to the "higher" affiliations. Religious instruction is minimal. That domain is viewed as the purview of parents and ordained leaders, not of the young, unmarried women who largely serve as teachers. The role of nature study and science in the Amish schools must be understood in this broader context.

But why inquire about the role of nature study and science in Amish schooling at all? One reason is that science is an important source of the knowledge and understanding needed to appreciate environmental changes that are affecting vegetation, wildlife, agriculture, and climate across the globe. According to the late rural sociologist Frederick Buttel and his colleague Peter Taylor, "We know we have global environmental problems because, in short, science documents the existing situation and ever tightens its predictions of future changes."[8] Also, the Amish themselves increasingly embrace cutting-edge science to solve problems they face in business and in personal affairs. Health-care diagnoses and treatments for the rare heritable disorders that disproportionally show up in Amish communities require a basic understanding of genetics, as do agriculture and animal husbandry. A foundation in the scientific logic underlying chemistry and biology, especially the importance of evidence based on experimental tests, can be valuable for sorting through testimonials and claims about nutritional supplements. It's worth asking what eight years of parochial schooling provide for the Amish in the way of scientific understanding and whether this foundation gives them the tools they need to understand ecological processes that affect the well-being of their community.

Surprisingly, we know little about the role of science education in parochial schools, perhaps because most researchers take its absence from the formal curriculum at face value. The two major books published on Amish education, for example, do not mention science at all.[9] A short 1998 article by Marlow Ediger does address the topic and notes that science, if taught in the Amish schools, typically involves searching for a correct answer and memorizing facts, like parts of a grasshopper.[10] Cory Anderson's synthesis of works on Amish education takes a broader view and argues that the Amish, who see themselves as "suspended between Christ's first and second appearances," promote *wisdom*, based on humility and respect for authority, rather than *critical thinking and science*, which explore multiple, subjective viewpoints and offer a way to choose among them.[11]

During our interviews, Amish parents and teachers frequently asked us what science *is* and expressed confusion about the difference between science, nature observation, and "common sense." Asked to give his definition of science, one Old Order man closed his eyes, thought deeply, and replied, "Well, you're gonna make me look stupid. I guess science is the theory behind what makes stuff work and go and all that kind of thing." The attempts by our interviewees to define science usually captured important elements of the scientific process, namely, the search for explanations of natural phenomena through careful observation. Their responses tended to downplay or omit other key elements of science, however, such as hypothesis testing, replicability, and falsifiability.[12] In other words, they did not emphasize a key element of science—that it is open to alternative explanations that are testable. Rather, in keeping with the unquestioning certainty of their religious beliefs, they were comfortable with science only when it was consistent with their religious ideas and followed established traditions. Within this framework, some leeway exists for the study of nature and mechanical processes, but attempts to move beyond them are usually frowned upon by the school committees.

Field Trips and Nature Study

Most Amish teachers describe the role of nature study in parochial schools as minimal. "Well, so nature is not, by default, a very big part of the Amish curriculum," noted one male Old Order teacher. "In all the years I've been going to the teachers' meetings, there's never been a topic about nature or animals," commented another Andy Weaver teacher. Yet probing more deeply inevitably turned up interesting examples of nature-related activities. "Field trips are pretty much expected," noted an Old Order teacher. In addition to picnics and camping trips, these may include visits to zoos, museums, robotic dairy farms, and myriad other locations. Teachers who organize and lead nature hikes often tie them to a study theme or a class activity. One Ohio teacher introduced the concept of "biome" and then took students to a marsh, where they waded around and kept track of all the creatures they found, including tadpoles and leeches. Another teacher took the students to collect walnuts, after which they dried and cracked them and made a walnut cake for the eighth graders. Through such activities, teachers who have a personal interest in nature can play a large role in fostering that interest among pupils. Many Amish birders, for instance, attribute their interest to a schoolteacher who asked them to keep a list of birds seen in the schoolyard and took them on nature hikes.

One unique way that Amish schoolchildren sometimes engage with nature is through pest hunts, in which pupils compete within or between schools to see who can kill the most pests. Teachers assign points on a sliding scale, with spiders and sparrows at one end, mice and rats in the middle, and groundhogs, raccoons, and larger animals at the other end, and students must bring in proof of their conquests. "Oh, yes," one Andy Weaver teacher said, "they love pest hunts." When asked, "Even the girls?," she replied, "Yes, the girls go after the spiders. They put in the most points. I didn't want to put spiders on the list, but that's where they got their points. They collected hundreds of them." She considered this especially significant, since "Amish ladies usually hate spiders and despise them spinning webs in their houses." Other teachers were more lukewarm toward pest hunts, noting that they can "keep the kids up too late at night" and that "it needs to have some supervision." One teacher in the Arthur, Illinois, settlement remembered, "We had to bring proof of sparrows, we took their heads, and mice and rats, we took their tails. Then we decided since we're working on honesty, we trust your word. If you say you killed ten mice, we believe you." Pest hunts strongly reinforce the ideas that humans have control over other living things and some of God's creatures are beneficial to humans, while others are not.

One example of the cultural parameters surrounding school-based nature study is an initiative by a local Amish businessman to put a "nature library" in parochial schools in Ohio's Holmes County settlement. Ultimately adopted by roughly half of the settlement's two hundred-plus schools, the project delivered a sturdy wooden bookshelf and nearly twenty reference books free of charge to schools whose teachers requested them. The curricular materials included field guides to birds, trees, and plants and some nature books published by A Beka, a textbook company for Christian homeschoolers. Though the ostensible goal of the project was "more nature awareness in the schools," as one teacher put it, a conversation with the project's coordinator revealed a more nuanced version. "Well, I don't know that I thought there was a deficit in nature awareness, I just wanted to be able to make it available." He continued: "It's the competitive attitude that we don't want. A lot of the Amish are involved in sports. I do not support it. Tournaments, I don't support tournaments. I think what the Bible teaches me is that I give and expect nothing in return. If we have to entertain in order to get people to contribute, something is out of balance." In other words, the goal of the nature-library project was less to foster ecological awareness than to promote nature study and enjoyment *as a healthy alternative* to the sports and entertainment mentality seen to be creeping in among Amish youth.

The project was not without resistance, however, and eventually the Amish Education Advisory Committee recommended that teachers not make the nature library their main curricular focus. Reaction against the project took two forms. First, some parents felt that too much emphasis on nature study took away from the parochial schools' main focus, core subjects. One critic reflected, "They lose sight of the basics being pushed back when extras are added. It seems the teacher thinks every week we have to have something exciting going on." The more serious line of resistance was a reaction to the fact that among the reference books in the nature library was a two-volume set titled *Character Sketches*, which uses a Christian framework to associate animals with human character traits deemed desirable in God's eyes. A chapter on the honeybee not only covers the behavior and natural history of bees but includes a scripture lesson on the importance of "being a reliable messenger to those I am serving."[13] As it turned out, the *Character First* texts provided an opening for teachers who wanted to promote a controversial set of religious convictions, including the idea of "assurance of salvation." These teachers ran up against a strong sentiment to keep religion out of the parochial schools. A teacher who was censored and ultimately left the Amish reflected: "Well, that is the bottom line of it because we were making those connections from religion. . . . I mean you'd think that would be a parent's ideal. But just, quite frankly, there are Amish people who aren't 'born again' and who are scared about that, and so when their children became 'born again' at the age of fourteen, fifteen, eighteen, sixteen, whatever, this suddenly created a problem." Ironically, then, a project like the nature library, which on the surface promoted awareness of the natural world, got caught up in an internal debate about what the core curriculum should include and the extent to which teachers should be allowed to support, augment, or undermine the religious authority of parents and church leaders in their classrooms.

Science Lite in the Parochial Schools

The examples of nature study mentioned above are less likely to occur in the Swartzentruber schools, where nature is apt to show up in the form of wood-burning stoves in the winter or outhouses in the schoolyard. With a streamlined curriculum of phonics, spelling, reading, arithmetic, and, beginning in fourth grade, "Bible German," Swartzentruber schools "provide enough book learning for children to earn a living," while also "separating them from the dominant society."[14] Parents are rarely involved in school activities, and most school boards meet infrequently. Nature study, much less science, has virtually no role in this context. In contrast, Old Order schools in small settlements have an expanded curriculum that usually includes art, health, geography, and history and allows teachers more flexibility in constructing daily schedules.[15] Schools in the larger midwestern settlements offer even more opportunities for field trips, parental involvement, and supplemental curricular materials.

A handful of the teachers we talked with in less conservative and less isolated schools went beyond field trips and basic identification of plants and animals to explore what one Amish teacher called "science lite," which he defined as the study of "what makes stuff work and go." Teachers who supported these occasional forays into science lite insisted that they were acceptable if framed in the right way. "Let's study God's world as He made it," insisted one New Order teacher. Asked where the line was between nature study and science, one teacher replied, "There's no difference to me," while another pointed out, "Some Amish are afraid just because you call something science, they're automatically, 'Well, we want to stay away from science.' No, no, no, science is simply to me a study of nature." One Old Order man, for example, related that his son studied how magnets work and how you can know the age of a tree when you cut it up. "Stuff like that, yes, that is science but it's low-key science. It's not hard-core. We're all for that kind of science," he concluded.

Teachers interested in science lite usually have to put together their own curricular materials from public-library books or personal purchases since school boards rarely sanction science textbooks of any kind. "It would be really helpful to have a set curriculum" so that I "wouldn't have to invent it," complained one teacher. Sometimes they turn to sympathetic parents for help. "Last year we were studying the heart, and one of the board members works in a butcher shop," a teacher in Illinois told us. "And he had two pig hearts, and we cut it apart and looked." We also learned of teachers who brought a microscope to school and asked students to look at drops of blood or other microorganisms.

In one unusual but revealing case, a retired public-school science teacher has spent the last ten years visiting more than a hundred schools in Indiana and Ohio to perform hands-on experiments on topics such as refraction of light, sound pitches, and Newton's Three Laws of Motion. A favorite experiment of students is the Cartesian-diver experiment, in which students manipulate the buoyancy of a figurine diver and make it go through an underwater basketball hoop. According to an Amish friend, "The children always love having him come. He's smart enough not to call it 'science' or 'experiment.' It will be, 'I'm going to show you something really neat.' And the children have no idea that they've just learned a lesson on volume displacement, or air density, or whatever." Interestingly, this teacher's motivation did not come from a perception that the Amish were deficient in science instruction; rather, he and his wife had become enamored with Amish parochial schools and wanted to help out. The acceptance of his "topics" rests on his pragmatic approach and his reputation as a Christian who will not challenge core values. "I won't do anything on evolution, and I don't get much into living things," he told us. "I'm very careful on that."

Crossing the Line: Secular Science

Almost all Amish draw a sharp line between nature study or science lite, on the one hand, and what several of our interviewees called "secular science," on the other.

Asked "what comes to mind when you think of a scientist," most of our interviewees described men in white coats working on experiments with mice, often for personal gain. One Old Order schoolteacher said, "If you mention science to the Amishman, he's going to see the guy in a lab looking at what he can next do that's against nature and is going to be harmful in the long-term. [He's] going to see science as, 'those are the people that gave us stem cell abortion [and] chemicals.' . . . Science has got a totally negative connotation for the Amish." In addition, most Amish associate science with liberalism, elitism, and immorality. "When I think of a true scientist," an Old Order man told us candidly, "I kind of think of an ungodly life."

Curricular options do exist for Amish teachers or school committees who are interested in science presented in a biblical frame, and some veteran teachers say that they are seeing more and more Amish schools in progressive settlements adopt them. Rod and Staff, a Kentucky-based conservative Mennonite publishing company, puts out the *God's World* science series for grades 4–8.[16] The grade 4 textbook, for example, includes detailed material on types of precipitation, kinds of circuits, and many other topics covered in public-school science curricula. However, it also has chapters titled "God Gives Us Weather" and "God Made Light," and it makes claims such as "God made birds fly so they could escape from their enemies." Such a biblical framing of scientific facts does not meet the standards of science, because it attributes natural outcomes to the intelligent actions of a deity, which cannot be tested empirically. As a result, these texts do not qualify as meeting state-mandated science requirements for the public schools in any state where the Amish have settled. Yet even these textbooks that present science from a Christian perspective are unacceptable to most Amish school committees.

A science curriculum has difficulty gaining traction in Amish schools for several reasons. The Rod and Staff science textbooks are controversial in part because they come from a Mennonite publisher and are seen to promote an evangelical stance with which many Amish are uncomfortable. In addition, Amish teachers themselves had no exposure to science as students, and thus on a practical level they do not feel prepared to teach the subject. Those few teachers who do promote science often have other differences of opinion that alienate them from their school boards, not to mention that as young, unmarried females, they are near the bottom of the status hierarchy. More broadly, most Amish school boards find very little in a science curriculum to be valuable for living an Amish life. To the contrary, they see much that would potentially threaten it, especially the concept of natural selection. "The reason we don't have science [in the schools] is Darwin," one Old Order bishop said pointedly. "You just can't talk about evolution if you want to live in this community."

The experiences of Amish teachers who have tried to introduce biblically framed science reveal just how deep the resistance lies. One Old Order man recounted that a request to adopt Rod and Staff science textbooks in his community's school was met with complete silence at a meeting of parents and the school committee, which meant a resounding thumbs-down. A young Andy Weaver teacher in Ohio recalled how the school committee reacted to her attempts: "I wanted to have science in school, but they were dead set against it. It was an evil word for them. Rockets, evolution, that's what they think. One board member, he was very old-fashioned. He thought the sun revolves around the earth, the sun was not a star. I just couldn't deal with those narrow-minded ways so I left [that school]. They practically called me un-Christian!" She lamented that if she had just called it "nature study," it would have been all right. A veteran teacher agreed and described how he had framed the study of rocks and minerals in his class. "I said, 'Let's see how many kinds of rocks we can find in the creek bank here, and let's see if we can figure out what they are.' If I'd called it 'geology,' it would not have gone."

The experience of a young man, formerly Amish, who grew up in the Rexford, Montana, Amish settlement further illustrates the line between science lite and secular science. "Science really interested me even back when I was fourteen or fifteen," he said, which had led him to read encyclopedias and science books at the local library. "One memory I have of grade school was when I told my friends that Jupiter has moons, the moon [we see] wasn't the only moon. I got roundly mocked for that." Now a physics major at a large university, he reflected on the Amish resistance to science:

> I think it's just the scientific method, the idea that you can use evidence and even statistics to form reasonable conclusions, as opposed to just going with, say, an anecdote. And they're generally anti-science, not just uneducated in science. So that makes it really difficult to approach them with a scientific argument. Because it's science that tells them that the universe is fourteen billion years old, and, obviously, all Amish would believe in a young earth. If science is *that* wrong, then everything about science must be wrong.

Most Amish school committees, then, keep science at arm's length because they realize that evolution threatens the very foundation of their biblically inspired view of the origins of humankind.

In the Amish view, ultimately the Bible itself is the sole evidence that counts, and science can therefore only have a role if the evidence on which it is based fits with the Bible. One Old Order Amish man summed up the Amish view this way:

"For the most part, we believe science can all be boiled down to the creation of the earth. Science is God's perfect design. . . . We're a simple people, chased out of our homeland because of religious beliefs. We grow up with the idea that we're just pilgrims traveling through here, and we don't make it our business to have to figure out how things work. We do what we've always done." The Amish thus share with many other creationists a tendency to sidestep the debate over evidence about evolution. "We definitely don't buy the millions and millions of years," one school-committee member explained, "because we believe scientific evidence that is biblical, and that fits with our understanding of the world."

Nature and Science in Amish Childhood

Why do some people grow up to be nature enthusiasts, actively concerned about the welfare of plants and animals, while others remain indifferent or even callous toward the natural world? The easy answer is that childhood exposure—being outdoors, taking care of animals and participating in outdoor activities—predisposes people to like nature and to work for its protection. A close look at Amish childhood reveals a more nuanced explanation, however. At a time when urbanization and market economies are everywhere transforming the relationship between humans and the natural world, the Amish preserve a childhood spent outdoors, close to animals, and relatively unmediated by technology. This is a considerable accomplishment, one that can give the Amish a deep familiarity with natural processes. However, contrary to what environmentalists might imagine, the biblically inspired view of nature they acquire does not automatically translate to "love of nature" or "nature conservation." Even if they find nature appealing, Amish youth are exposed to a largely unsentimental and utilitarian view of nature, and they learn to categorize plants and animals into those that are beneficial to humans and those that are not.

The implications of the lack of "secular science" in Amish parochial schools are similarly complex. Though no studies exist that compare science achievement among Amish and non-Amish pupils,[17] it's worth remembering that science-literacy rates are relatively low even in the general population.[18] Perhaps the Amish acquire all the "science" they need to know through the experiential, community-based, and lifelong learning that has served them so well in their small businesses. Many Amish leaders continue to believe that formal science instruction in the schools offers little of importance for their lives. Amish who do see value in critical thinking and scientific reasoning must therefore acquire these skills in other contexts. Nor does the official stance against science instruction in the parochial schools equate with a prohibition on using the fruits of scientific inquiry. Many Amish selectively and creatively use the achievements of science in agriculture, animal husbandry, health care, and other areas. Their limited exposure to the scientific process in the parochial schools, however, means that they do not have a strong foundation for understanding regional and global ecological processes, particularly those that are largely invisible in their daily experience. Science is not a series of piecemeal facts but involves the ability to move from small observations to their implications on a wider scale. This scientific logic of looking for patterns and teasing out the connections between the micro and macro levels is also fundamental to the analysis of household resource use and the carbon footprint, the topic of our next chapter.

CHAPTER THREE

 The Amish Ecological Footprint

An Amish horse and buggy makes its way along the shoulder of a major roadway while cars, trucks, and eighteen-wheelers fly past at highway speed, the vacuum of their passage rocking the fragile conveyance. A more potent metaphor for the contrast in material culture between the Amish and their modern neighbors could hardly be imagined. The iconic buggy is shorthand for a suite of "old-fashioned" practices that non-Amish observers assume lie behind that buggy and its passengers. The Amish intrigue us partly because they seem like a historical anachronism.[1] Furthermore, the conveniences the Amish forgo are some of the very things critics of modernity tell us are the material causes of our current moral, social, and environmental crises.[2] The Amish seemingly avoid the compulsive consumption of a "throw-away" society, and they largely shun the machines, devices, and behaviors that pollute, generate waste, use scarce resources, and fuel the consumer society. For environmentally aware North Americans who are attentive to the stresses humans place on natural systems, the Amish seem to practice a way of life that is more frugal and sustainable in a world with finite resources.

Amish and non-Amish alike are aware that the consumer choices we make have, at the very least, economic consequences. Blueberries purchased in Pittsburgh in November are expensive because they are shipped by air from Chile or Argentina. The monetary and environmental costs of out-of-season blueberries come from the inputs that occur at each step in the "life cycle" of the product.[3] These include the land, labor, fertilizer, water, and fossil fuels needed to grow and harvest the berries, the infrastructure of intercontinental trading partnerships, and the labor and fuel to fly the berries to the United States and deliver them in a refrigerated truck to your grocery store. Even at the grocery store, costs include paying employees, heating and lighting the store, and upkeep of the building. The environmental costs of things we buy or use come from the energy and materials needed to harvest or make them, the costs of moving them around the world, and the costs of disposing of the wastes generated when their useful life is done.

From this perspective, everything we do has environmental consequences. Growing food, discarding batteries, spreading manure, and buying a computer all carry resource and energy costs. And while some resources, like fresh water and oxygen, are renewable within a human life span, others are finite. Unrenewable resources, like oil, iron, copper, or gold, require huge amounts of energy and labor to find and harvest them and often carry a high environmental cost. We seldom think about where the "stuff" in our lives really comes from, but scientists would argue that only when we consider the backstory can we really understand our impact on the planet.[4] As the human population has grown, even trivial differences in resource use are magnified by billions of consumers. The effort to understand and quantify the environmental effects of our choices has given rise to the concept of the "ecological footprint," the idea that the way we live leaves a mark on the earth.[5]

In its original sense, an ecological footprint attempts to quantify the physical area, in acres, needed to generate all the goods and products a household or a country uses and to dispose of all the wastes produced. Behind this analysis lies the larger question: Is the way we live exceeding our planet's resources? A family who owns two homes and four cars, flies thousands of miles annually, purchases and discards clothing according to the whims of fashion, and boasts rooms full of equipment for sports, recreation, and entertainment has a large ecological footprint. Conversely, an Amish family that owns two buggy horses, travels six thousand miles annually in a hired van, makes and passes on much of its plain clothing, and lacks television, radio, and internet technology would have a smaller footprint. Non-Amish who worry about their environmental footprint are motivated by a desire to live more sustainably and to use a smaller, more equitable fraction of the earth's finite resources. But in the end, the size of your ecological footprint depends only on the choices you make, not the reasons why you make them. Outsiders may assume that the Amish live as they do in an attempt to conserve the earth's resources. Even if Amish home economy is not driven by environmental motives, however, the Amish may in fact provide insights for those trying to live within planetary limits.

Many writers have speculated about the differences between Amish and non-Amish resource use, but their analyses have been based on small sample sizes, sometimes on only one or two families.[6] We wanted to devise a survey that compared the ecological footprint of Amish and English households on a broader scale. We needed a survey that would strike a balance between asking for sufficient detail and being too demanding. If Amish and English footprints differed, then in what aspects of their lives would these differences be evident? In this chapter, we report the results of a comparison of Amish and non-Amish households in the rural and small-town areas of northeastern Ohio. We compare their home economies and try to assess where they fit on a spectrum of earth-friendly livelihoods espoused by the larger environmental community.

Household Livelihoods and Their Environmental Implications

In order to examine Amish livelihoods, we drew from our fieldwork and interviews with Amish families. We also conducted a written survey about household practices, diets, and purchases. In order to compare things like travel, home heating, gardening practices, and purchasing, it was important to survey Amish and non-Amish living in the same geographic and economic region. All surveyed households experienced the same climate and weather, had similar options for local travel, were a comparable distance from larger urban centers, and had access to a similar landscape for recreation, gardening, and resource use. We queried Amish and non-Amish residents of Wayne, Holmes, Tuscarawas, and Coshocton counties. We sent out 1,500 surveys and received a total of 377 responses—154 from Amish and 213 from English households, along with the results of interviews with 10 Swartzentruber families, which brought the total Amish responses to 164.[7] All the Amish households were from populated rural settings; the English respondents were a mixture of rural residents and residents of small towns and villages. Amish respondents were members of the four main affiliations in the Holmes County settlement—the Swartzentruber, Andy Weaver, Old Order, and New Order groups—who interact with the larger economy in somewhat different ways.

One obvious difference between Amish and non-Amish households that affects resource use is family size. The demographic profile of the Amish community in the Holmes County settlement is dominated by children and by parents with large families. In our random sample, 70 percent or more of the four Amish groups had households with five or more members, as table 3.1 shows. Only 9 percent of English families in our data set had five or more members. Conversely, two-thirds of our English respondents were households of one or two people, but there were few one- or two-person Amish households. Swartentruber and Andy Weaver households had significantly larger families than Old Order and New Order Amish.[8] All four Amish groups had significantly larger families than our English respondents.[9]

We divide our survey results into three broad categories, beginning with the environmental consequences of transport and travel. Next, we focus on basic home infrastructure, especially energy use. Finally, we examine patterns of consumption, including both diet and purchases of goods and services. We then translate these categories into an ecological footprint that is relevant to a global world-view. In many

TABLE 3.1.
Mean family size reported by groups in our survey

Affiliation	*n*	Mean family size (SE)	Households of 5+ people (%)	Households of 8+ people (%)
Swartzentruber	10	8.40[a] (1.529)	70	60
Andy Weaver	41	7.71[a] (0.419)	88	51
Old Order	76	6.05[b] (0.307)	71	28
New Order	37	6.24[b] (0.347)	81	24
English	213	2.66[c] (0.090)	9	< 1

Note: A Welch's robust analysis of variance showed significant differences in mean family size among affiliations ($F_{(4, 47.054)} = 77.902, p < 0.000$). Means having the same superscript are not statistically different from one another (Tukey's test).

respects, compared with their English neighbors, the Amish are models of what it means to "act locally." For example, Amish self-sufficiency reduces the "food miles" that lead to their dinner tables, making their diets more energy efficient. In this and other ways, the Amish circumvent the global consumer network, converting their own labor into things they need in their everyday lives. Some aspects of the home economy, however, are intrinsically tied to the larger economic system. Whether a home is on or off the grid, burning fossil fuels is the easiest way to generate heat and light, and in their energy use the Amish and the English are more alike.

Transportation and Travel

Mobility is a hallmark of modern America, and the idea that the Amish live happily without automobiles piques the curiosity of outsiders. The English move about in a cloud of gaseous and solid pollutants that are emitted from the tailpipes of their mostly fossil-fuel-burning vehicles. In our survey, English households owned, on average, 2.08 automobiles each, with an average gas mileage of 23.47 miles per gallon.[10] The Amish do not own cars, but they do ride in them. How do their modes of transportation compare with those of their English neighbors? We asked our respondents to estimate the miles they traveled annually by buggy, van, car, train, bus, and airplane. Table 3.2 summarizes their annual miles traveled by buggy and hired van. Not surprisingly, English respondents far exceeded the Amish in miles in a private vehicle. Only two Amish families reported private automobile miles, perhaps from rumspringa-aged youth. The average mileage driven annually by

TABLE 3.2.
Family vehicle travel reported by groups in our survey (miles per year)

Affiliation	*n*	Mean hired van transport for long trips (SE)	Mean local van transport (SE)	Mean buggy distance driven (SE)	Mean private vehicle transport (SE)
Swartzentruber	10	97.60 (36.733)	220.8 (76.361)	2,509.20 (658.353)	0.0
Andy Weaver	41	1,313.75 (367.481)	4,479.51 (1,273.354)	1,406.34 (346.719)	0.0
Old Order	71	2,077.18 (411.759)	5,168.79 (920.578)	1,097.92 (149.170)	478.87 (414.017)
New Order	37	1,761.62 (334.203)	2,803.78 (561.800)	878.92 (162.499)	0.0
English	213	1.29 (0.944)	15.36 (11.005)	0.0	21,357.73 (973.963)

English families was 21,357 miles, and half the English we surveyed drove more than 18,000 miles a year. In addition to the environmental costs of using fossil fuels, it also takes resources and energy to manufacture cars, transport them from the factory to the distributor, and eventually to dispose of them (and their many tires).

Among the Amish groups, only the Swartzentrubers have strong church restrictions on riding in cars and vans, as is evident in table 3.2. The Swartzentrubers had approximately eight times as many buggy miles per year as van miles. Conversely, Andy Weaver, Old Order, and New Order Amish households reported approximately four, seven, and five times as many van miles as buggy miles, respectively. Vans may carry more passengers, but they also typically get lower gas mileage than private vehicles. Amish families still have a far lower annual vehicle usage than English families, but there is considerable variation among families in all groups. Among the Amish, 15.8 percent of households (26 families) reported traveling more than ten thousand miles in vans annually. Conversely, 18.3 percent of English households (39 families) drove less than ten thousand miles in a year.

Buggy travel is not without environmental impacts, however. Buggies, like automobiles, must be fabricated from metal, fabric, and other materials. In addition, horses have environmental footprints in the same way that humans do. While horse transport in itself may be fossil-fuel neutral,[11] owning a buggy horse has environmental impacts.[12] Unlike automobile travel, in which emissions are produced only when the car is being driven, horse travel has a continuous cost; you cannot "turn off" the horse. It requires feeding and maintenance all year long. Horses pastured and fed on an Amish farm consume hay grown with a limited fossil-fuel input. But

TABLE 3.3.
Annual family public-transport travel reported by groups in our survey (miles per year)

Affiliation	n	Bus travel (SE)	Train travel (SE)	Air travel (SE)
Swartzentruber	10	662.60 (252.094)	923.20 (271.824)	0.0
Andy Weaver	41	42.50 (30.605)	407.50 (265.845)	0.0
Old Order	71	481.27 (108.069)	476.06 (361.516)	415.49 (242.801)
New Order	37	271.62 (123.003)	89.19 (51.375)	2,670.27 (680.706)
English	212	132.99 (56.445)	18.31 (13.589)	3,527.27 (495.516)

if the hay is purchased from a non-Amish farmer, the farming itself involved fossil fuels. Buggy horses must be reshod every six to eight weeks, requiring materials and energy at the farrier's. Greenhouse gases are generated when horses digest food (enteric emissions) and in management of their manure. While the manure from horses (and other farm animals) is often applied to fields as fertilizer, decomposition as it composts and degrades produces carbon dioxide, methane, and nitrous oxide.[13]

Both Amish and English families also used public transport, including buses, trains, and airplanes (table 3.3). Like private transportation, public transportation has environmental costs, from resources used to build the train or airplane to the carbon generated by burning fuel. Swartzentruber Amish were the largest users of both bus and train transport, typically for long-distance travel to family events or for health care.[14] Other Amish groups reported considerable travel by hired van, preferring the more flexible option of hiring a driver for longer trips. New Order Amish were low in bus and train use but reported considerable air travel, a mode of transport that is permissible for other Amish affiliations only in emergencies. Differences among the groups highlight the degree to which Amish families travel, a topic we consider in chapter 9. While airplanes use a notoriously high amount of fuel, all forms of public transport carry many passengers at a time, making them less environmentally costly. In our study area, English respondents reported a somewhat higher airplane mileage than New Order Amish, but for urban English populations, air travel might represent a significantly larger part of their environmental footprint.

Household Energy Use

Household energy use is one of the largest components of our environmental impact, and most of that energy comes from a diffuse electric power grid incorporating

Unlike other groups, the New Order Amish allow travel by airplane, which adds an additional component to their carbon footprint. Photo by Barry Stein

coal, natural-gas, nuclear, biomass, wind, solar, and hydroelectric power production. All English households in our survey were on the commercial power grid, using electricity for lighting and sometimes for heating. But Amish homes varied in their lighting sources. Swartzentruber Amish used only kerosene lanterns and battery flashlights or battery lanterns. In addition to these sources, Andy Weaver households also used clear (white) gas and natural gas for lighting. Among Old Order families, piped natural gas, along with natural gas, propane, and battery lamps, were the principal lighting sources, but more than half of Old Order families also used solar electricity for lights or other devices or to charge batteries for lamps. One Old Order household even reported using microhydro power for lighting. The thirty-seven New Order families in the survey were the most eclectic in their use of lighting devices. In addition to natural-gas lighting, most used battery flashlights, eight homes had grid electric power, fourteen had solar electricity, and several used propane lamps or reported using wind or generator electricity.[15]

Solar energy is widely touted as an environmentally friendly source of electrical power. But with ubiquitous access to the electric grid in densely populated northern Ohio, English homeowners must make a substantial capital investment to change to solar power. Only two English homes in our survey (1 percent) reported having solar panels, while 38 percent of New Order and 61 percent of Old Order Amish said they used solar panels for electrical power in their homes.[16] These Amish groups embrace solar technology, not necessarily because it is "green" but because it is a permissible option for generating current in their homes. Typically, Amish homes use solar panels for recharging 120-volt batteries, running LED lights or small pumps, or powering sewing machines or small appliances. Photovoltaics have been deemed acceptable up to certain limits by Amish church districts, even though they potentially open the door to other electric appliances that have long been avoided by the Amish. Their growing use of solar power is a good example of the Amish ability to make creative use of modern technology within the constraints of their local Ordnung.

Home heating and cooling are typically a large component of the household footprint, and different fuel systems have different environmental implications. Historically, the Amish heated their homes with wood, although in the Holmes County settlement today more liberal Amish families use natural gas as their principal fuel. But in this relatively rural region of Ohio, wood was widely used by Amish and English alike. Swartzentruber families burned exclusively wood; among other Amish households, 65–70 percent burned at least some wood. Of English respondents, nearly 27 percent said they burned some wood as a fuel, although only five families reported using primarily wood for heating. A few Amish respondents burned coal, and English families also used electricity, propane, natural gas, or heating oil as their heat source.

Wood and coal are inexpensive, and wood is sometimes free. But as fuels, both wood and coal come with high environmental costs because they produce large amounts of greenhouse gases as well as particulates with potential health consequences. To be sure, wood lacks the production costs of petroleum-based fuels. If homeowners harvest it from their own woodlots or local area, the transport costs are lower too. And wood is sometimes described as "carbon neutral" because the carbon released upon burning is carbon dioxide, which trees incorporate by photosynthesis, and not "fossil" carbon.[17] But there is a lag of decades between the release of carbon from burning wood and its equivalent uptake into new forest growth.[18] In fact, land-use changes and burning of forests are major sources of global greenhouse emissions, and thus the carbon costs of burning wood cannot be ignored. Some Amish households use scrap lumber from local sawmills as firewood. Harvesting dead trees has the lowest environmental impact, while cutting live trees for firewood is less environmentally friendly because it removes vegetation that absorbs carbon dioxide and prematurely converts live wood into greenhouse gases. Despite its long tradition as a self-sufficient frontier fuel, then, wood is not without environmental costs.

Old Order and New Order Amish in the Holmes County settlement have used natural gas since the 1950s.[19] In our survey, 93 percent of Old Order and 92 percent of New Order households said they had natural-gas lines. In 2014, bishops in the Andy Weaver community approved natural-gas heat and light for all members of their congregations, creating an avalanche of orders for installations with the local gas provider. In the summer of 2015, 70 percent of our Andy Weaver respondents reported having natural gas in their homes. Environmentally, natural gas has a lower carbon output than other combustible fuels, but like all petroleum products, it requires drilling, pumping, processing or refining, and transporting by pipeline or carrier. Because the Holmes County settlement is home to significant gas and oil exploration, some rural English and Amish have gas wells on their property and thus have a predetermined quota of "free gas." Those homes have no internal metering, but if they use more than the quota, they are charged for additional usage. The quota for free gas is usually around 250,000 cubic feet, which is well beyond the amount normally given as "typical home usage." If someone with free gas told us what they paid for natural gas, we assumed this was for usage over the quota.

Natural-gas use by groups in our survey is shown in figure 3.1. We were surprised by the very high usage among Amish families, even those households without free gas. The average English household used 95,000 cubic feet, very close to what is typical for a single-family detached home. All three of the higher Amish groups used more, sometimes much more, natural gas than their English counterparts, contributing to the high Amish environmental footprint for household heating. The weighted average gas use across the Amish affiliations was 149,500 cubic feet per year, 57 percent higher than the English average.

Home size is a crucial variable in determining the energy required to heat a house. As table 3.4 demonstrates, English families have significantly smaller homes than the Swartzentruber, Andy Weaver, and Old Order Amish. Amish homes are usually made of wood, and many older homes may lack the insulation that a modern dwelling would have. Maintaining a comfortable living temperature in a large, poorly insulated home requires a lot of fuel, which helps to explain Amish natural-gas use. Homes with thermostats reported their target winter home temperatures on our survey. As shown in table 3.4, the Andy Weaver Amish kept their homes

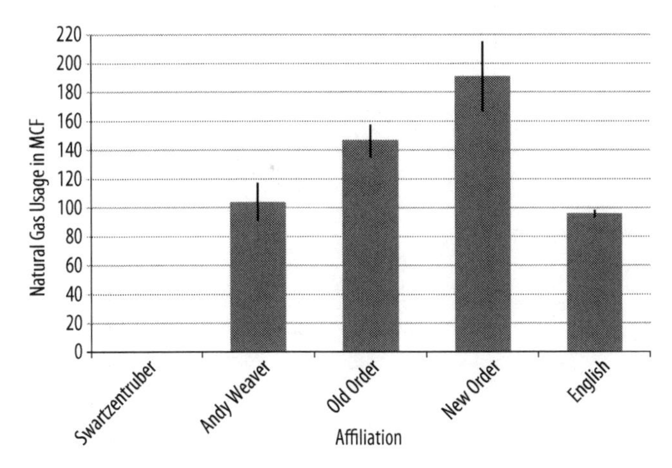

Figure 3.1. Average natural-gas use in thousands of cubic feet (MCF) by Amish and non-Amish households in our survey who reported using natural gas as a part of their home energy mix. The error bars represent standard error of the mean. By Marilyn Loveless

at an average temperature of 73.3 degrees Fahrenheit, significantly higher than the mean temperature of around 70 degrees reported by other groups. Differences in the acceptable home temperature would affect energy use by individual households.

Amish homeowners may also differ from English families in their perception of home economy. The Amish are frugal in many realms of their lives, but the convenience of gas heat, when compared with historical alternatives like coal and wood, may seem like a bargain even if their annual natural-gas bill is two thousand dollars. This mentality is especially common in households with natural-gas wells on their property. As one New Order minister told us before we even conducted our survey, "You can be guaranteed that people with free gas use more than those who pay for their gas." The unintended outcome is a larger environmental footprint for home energy use. While that might be a motivating factor for energy conservation among some homeowners, it is not a major consideration for many families, who are more likely to think in terms of cost, convenience, and comfort.

You Are What You Eat

Among the values most closely associated with the Amish are simplicity, frugality, and modesty. Outsiders tend to think of the Amish in terms not of what they own but of what they do without. Amish homes are imagined as being Spartan, and this leads us to expect the consumer profiles of these households to differ from those of non-Amish households. Outsiders also make many assumptions about the Amish

TABLE 3.4.
House size and preferred indoor temperature reported by groups in our survey

	House area (sq ft)		Thermostat temperature (°F)	
Affiliation	*n*	Mean (SE)	*n*	Mean (SE)
Swartzentruber	9	3,678.22[a] (487.667)	—	—
Andy Weaver	31	4,095.45[a] (258.516)	16	73.3[a] (0.598)
Old Order	60	3,823.68[a] (193.792)	49	71.20[b] (0.394)
New Order	29	3,587.97[ab] (194.190)	22	70.43[b] (0.419)
English	175	2,580.69[b] (99.135)	208	69.43[b] (0.211)

Note: Analysis of variance showed significant differences in house area ($F_{(4, 299)}$ = 16.440, $p < 0.000$) and in mean thermostat temperature ($F_{(3,297)}$ = 12.928, $p < 0.000$). Means having the same superscript are not statistically different from one another. Swartzentruber homes are heated by wood and have no thermostat.

diet, which some entrepreneurial Old Order women have answered by publishing cookbooks or running home businesses offering "meals in an Amish home" to tourists. We wanted to see what differences in diet and household spending might tell us about the role of consumption in Amish and non-Amish homes. To understand where the food on their tables came from, we asked our survey participants to record their eating patterns and tell us about gardening practices, wild-food gathering, patterns of sourcing their food, and meals eaten outside the home. Survey respondents estimated how many servings of different foods their families consumed daily. The average diet against which they measured their eating habits is given in table A.1. While this is a somewhat imprecise measure, it provided data we could enter into the footprint calculator and gave us some idea of how Amish and English families might differ in their eating habits.

Meat and dairy consumption are large elements of a dietary footprint because animal foods are energy and resource intensive. Animals consume foods (like grain) that can be eaten directly. Cows are also a major source of methane, which has strong negative effects on the global atmosphere. None of our Amish families were vegetarian, and surprisingly few English were.[20] On average, Swartzentruber, Old Order, and New Order Amish families ate more servings of meat per day than did Andy Weaver Amish and English households. All four Amish groups reported consuming more dairy products than did English households. English families, however, reported a

TABLE 3.5.
Egg consumption and meals delivered to home or eaten out reported by groups in our survey

	Eggs eaten per person per week		Meals delivered to the home per month		Meals eaten out per person per month	
Affiliation	*n*	Mean (SE)	*n*	Mean (SE)	*n*	Mean (SE)
Swartzentruber	10	12.68 (2.423)	10	0.07 (0.051)	10	0.63 (0.312)
Andy Weaver	40	5.58 (0.494)	41	0.52 (0.096)	40	0.80 (0.137)
Old Order	76	5.80 (0.354)	75	0.70 (0.101)	75	1.02 (0.111)
New Order	37	5.34 (0.373)	34	0.69 (0.141)	34	1.00 (0.116)
English	201	3.99 (0.190)	212	0.28 (0.068)	212	6.12 (0.413)

slightly higher consumption of snacks and beverages than Amish households. Of the Amish groups, the Swartzentrubers had a diet somewhat higher in meat and dairy and lower in vegetables and grains than other Amish groups. But despite the suspicion by non-Amish that mashed potatoes, noodles, and pie are consumed daily by every Amish, our data suggest that Amish and English households are similar in the kinds and quantities of foods they consume.

Several exceptions to this generalization are worth noting. First, based on our survey, per-person egg consumption was lower in English homes, higher among most Amish, and substantially higher among Swartzentruber families, as table 3.5 demonstrates. Many Amish families keep chickens, and eggs provide a staple in the Amish diet. Furthermore, Amish families do, indeed, order pizza and other take-out meals for delivery to their homes, and they do so about twice as often as our English respondents. In contrast, English households reported eating out at restaurants six times per month, compared with just one meal per month for Old and New Order Amish individuals and even fewer for the Andy Weaver and Swartzentruber Amish. These observations may reflect the differences in household size between Amish and English respondents. English couples without children may eat out more often for convenience, while Amish families with many children are less likely to head to a restaurant for dinner.

The Amish are virtuoso gardeners, and our results strongly verified that. Virtually every Amish family in our study had a garden, and those gardens were large, as shown in table 3.6. Swartzentruber gardens were almost seven times larger than other Amish gardens, on average. Among the English, gardens were much smaller,

All the Amish families we surveyed had vegetable gardens, and they were typically much larger than the gardens of non-Amish households. Photo by Marilyn Loveless

and only 89 families (44 percent) had vegetable gardens. Furthermore, Amish families reported eating from their gardens more often during the summer months than did English households. Amish households were also far more dedicated to food preservation for future use.

Canning is a cultural imperative for the Amish, and every Amish family in our survey, even those few without a garden, said they canned for later consumption. This was a major summer activity for Amish households, and the volume of food they preserved was impressive. Only half the English households (54 percent) reported canning, and in much lower volumes. Canning is especially important for Amish kitchens because it is the principal method of food preservation to which they have access. Swartzentruber and Andy Weaver homes lack mechanical refrigeration, using an ice box or a spring house. Old and New Order houses typically have refrigerators (powered by gas or propane) but limited freezer space. Though shared, off-site freezer barns are one solution, canned food in the basement is much more convenient.[21] While much of what women canned was from their garden, few Amish had orchards, so they typically purchased fruit to can—as did English

TABLE 3.6.
Garden size and contribution to diet reported by groups in our survey

Affiliation	n	Mean garden area in sq ft (SE)	Mean family meals per week from garden (SE)
Swartzentruber	10	23,972.22 (8,243.945)	17.10 (1.069)
Andy Weaver	40	3,722.20 (476.781)	13.27 (0.455)
Old Order	69	3,165.68 (369.210)	13.60 (0.407)
New Order	33	3,491.48 (429.151)	12.81 (0.669)
English	89	1,355.61 (322.831)	8.67 (0.545)

Note: Calculations are only for those who reported having a garden.

Amish households in our survey were far more likely than non-Amish households to preserve food, and they canned vegetables and meat in much higher volumes. Photo by Doyle Yoder

TABLE 3.7.
Family sourcing of food items from noncommercial vendors reported by groups in our survey

	Average percentage raised at home or obtained from a neighbor					Fruit and vegetables canned annually (qt)		Meat canned annually (qt)	
Affiliation	n^1	Meat	Chicken	Eggs	Dairy	n^2	Mean (SE)	n^2	Mean (SE)
Swartzentruber	10	96.0	86.5	100.0	75.6	10	733.20 (138.505)	10	275.50 (44.786)
Andy Weaver	41	47.0	21.7	89.5	42.0	39	322.82 (26.876)	41	161.22 (13.310)
Old Order	76	67.5	30.0	79.5	52.2	73	320.22 (31.508)	63	65.21 (6.991)
New Order	37	71.9	16.6	62.7	43.5	37	299.86 (33.921)	26	38.92 (7.702)
English	212	15.4	3.3	18.5	2.8	112	58.82 (5.468)	16	18.19 (2.391)

n^1 = the number of households reporting in the survey
n^2 = the number of households who reported canning either fruits and vegetables or meat

respondents. In addition to vegetables, Amish women canned pizza sauce, tomato sauce, pickles, applesauce, cherries, and strawberries. Some even reported buying and canning pineapple.

Perhaps surprisingly, the Amish also canned meat—hamburger, meatballs, trail bologna, chickens, and venison. Canning is the cheapest, most effective way to preserve a large supply of a perishable resource. If you slaughter a cow, you have to prolong its usefulness. If your husband and four sons each get a deer, even trail bologna can't be kept fresh long enough to eat it all. Not surprisingly, Swartzentruber families are the champions in the canning competition, as table 3.7 shows, averaging 275 quarts of meat and 733 quarts of fruits and vegetables annually. With large families and restrictions on transportation and cooling technologies, their home economy depends on what they can grow and preserve to feed themselves. The other Amish groups averaged about half as many fruits and vegetables canned as the Swartzentruber but six times as many as the English. Andy Weaver and Swartzentruber families, without freezer options, canned more meat than Old and New Order families.

Amish families in our survey also produced a larger percentage of their food at home than did English families. We asked respondents to estimate the percentage of meat, chicken, eggs, and dairy products they purchased from a store, raised themselves, or obtained (by purchase or gift) from a neighbor or relative. We compared the degree to which Amish and English households obtained various foods locally

TABLE 3.8.
Percentage of families foraging for wild plants reported by groups in our survey

Affiliation	*n*	Mushrooms	Nuts	Wild berries	Greens
Swartzentruber	10	40.0	50.0	50.0	80.0
Andy Weaver	41	41.5	2.4	43.9	4.9
Old Order	76	42.1	9.2	27.6	5.3
New Order	38	32.4	2.7	37.8	5.4
English	213	16.9	4.7	25.4	3.3

or from the commercial food system. The results, shown in table 3.7, make it clear that Amish households are somewhat self-sufficient, at least compared with their English neighbors. Swartzentruber homes get most of their meat, eggs, and dairy products from their own farms or from their neighbors. Other Amish affiliations produce, on average, more than half of their eggs and almost half of their dairy products locally. Values vary more for meat and chicken, depending on individual circumstances. Although their case was unusual, one Old Order farmer told us that in fifty years of marriage he and his wife had never bought milk, eggs, or chicken at a store. Amish households reported eating more meat and dairy products than English families, but the fact that many of those products were produced locally, on farms with low fossil-fuel inputs, slightly reduces the environmental footprint that is typically associated with animal-based foods.

Responses about gathering wild foods further support the conclusion that Amish households rely less on the commercial food system and are attentive to elements of their natural surroundings. For example, Amish households were at least twice as likely as English families to gather wild mushrooms (table 3.8). Swartzentruber families outpaced all others in collecting wild berries, nuts, and greens. Amish households were also three to four times more likely to eat venison and three to six times more likely to eat local fish (table 3.9). Some English households engaged in foraging and hunting, but Amish families were more attuned to the edibility of their landscape. In contrast, grocery-store foods usually come not from local producers but from regional and national distributors, who source commodities from the least expensive and largest-volume markets and transport them many food miles to reach their destinations. Our data suggest that on average, Amish families are eating closer to home than English households. To the degree that this is true, the Amish diet has a somewhat smaller environmental footprint.[22]

On the other hand, Amish households are certainly not detached from the commercial food system. They buy many staples and pre-prepared items from the same

TABLE 3.9.
Percentage of families harvesting wild game for the dinner table reported by groups in our survey

Affiliation	*n*	Venison	Frog legs	Local fish	Lake Erie fish	Turtle	Rabbit	Squirrel	Water fowl
Swartzentruber	10	80.0	0.0	60.0	20.0	0.0	60.0	30.0	0.0
Andy Weaver	41	63.4	0.0	78.0	41.5	0.0	14.6	2.4	0.0
Old Order	76	64.5	6.6	48.7	39.5	0.0	19.7	11.8	0.0
New Order	38	56.8	8.1	37.8	24.3	0.0	13.5	2.7	0.0
English	213	19.2	0.4	9.9	8.0	1.8	4.2	3.3	1.9

grocery outlets that English families use. Amish children consume potato chips, soft drinks, and filled cupcakes just as English children do. One Old Order mother expressed concern about the processed and junk food showing up in the lunchpails of Amish scholars in recent years: "Pastries, pop, cheap cookies, Ho-Ho's, packaged sausage—they live cheaply, but they eat unhealthy," she said. Viewed from a different angle, the results presented in table 3.7 show that, apart from the Swartzentrubers, more than two-thirds of Amish families buy their chicken, and half or more get their dairy products from the store. These trends suggest marked changes in consumption patterns among Amish households that have come with the growth of shop culture.

The Stuff We Buy

We are surrounded by stuff. We buy things we need and want, and those things have costs behind them. A pair of shoes comes from a piece of leather, harvested, processed, designed, sewn, and shipped. That camo hunting jacket, your favorite teapot, or a bottle of shampoo—the production and disposal of all of these have costs. We tried to measure consumption by our survey groups in the same way that we assessed diet. We asked respondents to estimate their monthly expenditures in seven different categories, relative to an "average non-Amish American family." Average values came from the Cool Climate Network and were based on an average household's expenditures on goods and services.[23] We then measured what percentage of the "average" a family spent on each category. Because these categories were aggregated into groups of goods and whole clusters of services, they are not a very sensitive measure of consumption. We could not track exactly how much a family spent for shoes or for dietary supplements, but we could calculate group values and compare consumption in general terms.

In six of the seven consumption categories, both Amish and English respondents reported spending less than the national average. The exception was the category

"Organizations, charity, miscellaneous," on which both New Order Amish and English families said they spent more than the average. These results could reflect a systematic underestimate by families who want to think of themselves as frugal. Or respondents may have overlooked some items in their responses. The results also probably reflect the fact that our respondents all live in a relatively rural area with a low cost of living and may have modest family incomes. But even though most of these households felt they were living more frugally than the "average," we can still compare their spending relative to one another.

In five of the seven spending categories, the Swartzentruber Amish were distinct from all other Amish and English in having significantly lower levels of consumption. In only one category, "Vehicle and household maintenance," did Swartzentruber families match other Amish groups in consumption. Our data clearly show that the Swartzentruber Amish are far more frugal than other Amish affiliations. In three of the seven categories, English expenditures were significantly higher than those of any Amish affiliation,[24] but in four categories the levels of consumption of English households and more liberal Amish households were not statistically different.[25] Our data suggest that many liberal Amish households are not excessively frugal or that many English families are more frugal than expected—or both. In the Holmes County settlement, it seems that only the Swartzentrubers still meet outsiders' expectations of low consumption by the Amish.

Measuring the Amish Ecological Footprint

One benefit of a footprint analysis is that it integrates a wide array of personal behaviors and choices into a single measure, a score of the potential environmental impact of a specific way of life. Over the past twenty years, the concept of an ecological footprint has been adopted and modified by various organizations. Footprints can be calculated for individuals, households, cities, or nations. A comprehensive footprint analysis is extremely complex, and while protocols exist to guide such an analysis, the process requires far more data than we could assemble. But interactive resources, many of them online, allow individuals to estimate their resource use and calculate a simplified environmental footprint using only a few key indicators. We sought an interactive footprint calculator that was credible and detailed enough to be likely to distinguish among groups but could be implemented using information that we could reasonably collect from Amish and English households.

Online calculators are designed for non-Amish lifestyles. We needed one that could accommodate distinctive elements of Amish life. For our analysis tool, we chose the Cool Climate Calculator, developed by the Renewable and Appropriate

Energy Laboratory at the University of California, Berkeley.[26] This tool measures environmental impact at the household level and quantifies impact with a "carbon footprint," or an estimate of the carbon dioxide equivalents that result from various behaviors and resource uses. Because carbon is central to chemical reactions within living systems, it is a valuable proxy for processes by which we use and degrade natural materials.[27] Carbon is also a dependable measure of energy expenditure because most of the fuels that drive our modern economy are organic, often petroleum based, containing carbon. Scientists have identified carbon as a key element causing anthropogenic climate change, thus impacting the future of our planet. But even for those who are uncertain about the validity of that prediction, counting carbon molecules gives us a uniform measure for comparing the ecological impacts of Amish and non-Amish households on resource use.

Footprint-Survey Results

The Berkeley Cool Climate Calculator divides activities into the sectors we've considered: transportation, household economy, diet, and consumption. But instead of descriptive accounts of behavior, the footprint analysis summarizes those activities in the currency of carbon. Results for the four sectors are shown in table 3.10. Individual values are then summed to generate a total household carbon footprint. Family size is an important factor in the total carbon footprint, since big families typically consume, for instance, food and other goods and services, at a higher rate. As a result, comparing the carbon footprints of families of different sizes would disadvantage large households. To account for this, we compared carbon footprints of affiliations for large households (5 or more members) and medium-sized households (3–4 members).[28] Although there were many English homes with only two family members, few Amish households had just two members. Thus, we could not statistically test differences among these small households.

Consistent with expectations, the English transportation carbon footprint was much higher than that of Amish households. We included an estimate for van travel for Amish families, as well as a carbon cost for each buggy horse that an Amish family owned, but even so, the many miles driven by English families in private vehicles far surpassed the miles traveled by Amish households. The constraints imposed by the Amish way of life, in which public transport is used for long-distance travel, have a positive environmental outcome. In both large and medium-sized families, the English carbon footprint was between two and four times that of any Amish groups, but there were no significant differences among the Amish affiliations. Because of their limited van travel, we expected Swartzentruber households to have a very low

TABLE 3.10.

Components of household carbon footprint (average metric tons per year) calculated for groups in our survey

	Swartzentruber Mean (SE)	Andy Weaver Mean (SE)	Old Order Mean (SE)	New Order Mean (SE)	English Mean (SE)
	Large families (5+ people)				
	$n = 7$	$n = 36$	$n = 52$	$n = 29$	$n = 10$
Transportation carbon footprint ($p = 0.000$)	8.80[a] (1.166)	6.90[a] (0.9624)	8.26[a] (1.006)	7.57[a] (0.796)	23.510[b] (3.720)
Energy carbon footprint ($p = 0.125$)	35.76[a] (3.999)	42.64[a] (3.779)	46.49[a] (4.926)	42.19[a] (5.112)	27.58[a] (5.205)
Dietary carbon footprint ($p = 0.108$)	25.17[a] (3.370)	18.73[a] (1.255)	17.26[a] (0.836)	17.18[a] (1.375)	14.99[a] (1.265)
Consumption carbon footprint ($p = 0.001$)	2.06[a] (0.423)	7.60[b] (0.453)	7.98[b] (0.436)	7.31[b] (0.420)	8.03[b] (0.854)
	Medium-sized families (3–4 people)				
	$n = 2$	$n = 5$	$n = 11$	$n = 6$	$n = 35$
Transportation carbon footprint ($p = 0.000$)	3.85[a] (0.650)	9.00[ab] (3.973)	5.36[a] (0.682)	6.20[a] (1.636)	17.05[b] (1.518)
Energy carbon footprint ($p = 0.392$)	24.65[a] (3.950)	35.18[a] (7.244)	34.19[a] (4.108)	30.67[a] (9.699)	27.31[a] (2.429)
Dietary carbon footprint ($p = 0.040$)	7.00[a] (0.600)	7.74[a] (0.621)	10.30[a] (0.687)	11.78[a] (1.573)	9.25[a] (0.458)
Consumption carbon footprint ($p = 0.000$)	1.75[a] (0.350)	3.95[ab] (0.340)	6.25[b] (0.628)	11.33[c] (1.793)	10.44[c] (0.614)

Note: Means in the same row that have the same superscript are not statistically different from one another. Group means were tested using Kruskal-Wallis tests; Mann-Whitney U tests were used for post-hoc testing to verify statistical differences between groups.

travel footprint, and within the medium-sized households this was true. But in our small sample of Swartzentruber interviewees, several big families had large numbers of buggy horses, thus overriding the carbon benefits of their low vehicle usage. Within the other three Amish groups, the mix of van miles and horse carbon costs varied widely, and all three groups had similarly moderate transportation carbon footprints.

The story was different for carbon costs associated with household energy use, however. We factored in the carbon produced from burning wood, which affected both Amish and English homes in our survey. English households all used electric power from the grid, but Amish families burned much more natural gas, often for lighting as well as heating, and they typically heated larger home spaces. Furthermore, Amish households used other fuels that added to their carbon footprint: kerosene and/or white gas for lamps and diesel or gasoline for small motors. As a result, there were no statistically detectable differences in household carbon footprints either in large or medium-sized families between Amish groups or between the Amish and their English neighbors. Amish households, instead of being more environmentally frugal, incurred similar carbon costs to those of rural English households.

Although Amish households often procure and process their foods in different ways than non-Amish households do, our survey results did not show any differences in the average number of servings of meat, dairy, and other food categories—and the results of the footprint analysis reflected this. There were no significant differences between any of the affiliations or between Amish and English families in their dietary carbon footprints. However, we suspect that this failure to find differences among large families was a statistical artifact of our sample sizes, especially the few large English families in the survey. The online calculator for dietary carbon is directly affected by true family size. But no English family had more than eight members, while many Amish families were larger than that. We cannot assess precisely how this might affect the results, but we suspect that our statistical conclusions for dietary carbon footprints are not robust. Furthermore, many Amish produce a substantial portion of their food locally, a factor the online calculator did not take into account. These statistical and computational problems highlight the challenges of making comparisons between households with widely different home economies.

The consumption footprint from goods and services showed significant differences between affiliations in both large and medium-sized families. In both cases, Swartentruber Amish had very low carbon footprints, reflecting their limited in-

TABLE 3.11.
Average total household carbon footprint (metric tons per year) calculated for groups in our survey

	Swartzentruber		Andy Weaver		Old Order		New Order		English	
Household size	*n*	Mean (SE)	*n*	Mean (SE)	*n*	Mean (SE)	*n*	Mean (SE)	*n*	Mean (SE)
5 or more ($p = 0.903$)	7	71.79[a] (6.650)	36	75.87[a] (3.986)	52	79.99[a] (5.109)	29	74.25[a] (6.473)	10	74.13[a] (6.556)
3–4 ($p = 0.214$)	2	37.25[a] (4.350)	5	55.87[a] (7.829)	11	56.09[a] (4.263)	6	59.98[a] (10.895)	35	64.06[a] (3.583)
2	1		0		4	42.35[a] (7.987)	1		85	50.83[a] (2.006)

Note: Means in the same row that have the same superscript are not statistically different from one another. Group means were tested using Kruskal-Wallis or Mann-Whitney U tests; Mann-Whitney U tests were used for post-hoc testing to verify statistical differences between groups.

teractions with the larger economy. In large families, other Amish affiliations had similar consumption values to those of the English. For families with three or four members, however, the New Order Amish and the English had similar consumption levels, and the Andy Weaver and Old Order Amish shared a significantly lower consumption footprint. Consumption levels followed the trajectory from conservative to liberal Amish and modern English households.

Finally, we added together the four footprints for transportation, energy, diet, and consumption to obtain average total carbon footprints for large and medium-sized Amish and English families, as shown in table 3.11. Among large families, all five groups have carbon footprints of around seventy-five metric tons, and Amish and English households were statistically similar. In medium-sized families, the Swartzentruber had a numerically lower footprint, but we couldn't be confident that it was really different because the data for the Swartentruber were based on only two families. The sample size was just too small. There are, of course, differences in the particular values for each group, but the variation within the groups and the unequal sample sizes mean that none of the values are *statistically* significantly different from one another. Based on our data, we must conclude that Amish and English families have similar carbon impacts on their environment.

This is a surprising conclusion. Is it really possible that the massive transportation carbon footprint of the English is counterbalanced by the household carbon emissions of an Amish home? How is it that the Spartan lifestyle of the Swartzentruber Amish is not reflected in their carbon output? How can we explain these results?

One possibility is that the big Amish carbon footprints we calculated are in fact a direct result of the large families we are measuring. Dietary carbon, while not

TABLE 3.12.
Average total per capita carbon footprint (metric tons per year) calculated for groups in our survey

	Swartzentruber		Andy Weaver		Old Order		New Order		English	
Household size	*n*	Mean (SE)	*n*	Mean (SE)	*n*	Mean (SE)	*n*	Mean (SE)	*n*	Mean (SE)
5 or more ($p = 0.000$)	7	6.84[a] (0.481)	36	9.84[b] (0.809)	52	11.70[b] (0.981)	29	10.77[b] (0.724)	10	13.09[c] (0.670)
3–4 ($p = 0.255$)	2	12.45[a] (1.450)	5	16.80[a] (2.743)	11	15.16[a] (1.164)	6	16.93[a] (3.944)	35	17.77[a] (0.953)
2	1		0		4	21.18[a] (3.993)	1		85	25.44[a] (1.003)

Note: Means in the same row that have the same superscript are not statistically different from one another. Group means were tested using Kruskal-Wallis or Mann-Whitney U tests; Mann-Whitney U tests were used for post-hoc testing to verify statistical differences between groups.

as important as transportation or energy use, does contribute to the total carbon footprint. If a single person typically has an annual dietary carbon footprint of 2.8 metric tons of carbon dioxide equivalents,[29] then a Swartzentruber household of fifteen people will have a dietary carbon bill of more than 30 metric tons—more than enough to erase any frugality in other categories. Big families also need big houses; house size figures into the construction costs of a home and also affects the energy needed to heat it. So the high carbon footprints of Amish households may simply reflect their higher levels of resource use. When their carbon footprints are compared, Amish and large English families may simply not be very different.

To address this possibility, we did one final calculation. We computed the *per capita* carbon footprint for Amish and English individuals, dividing each household footprint by the number of people in the family (table 3.12). Among large families, the Swartzentruber Amish do have a statistically lower per capita carbon impact. The other three Amish affiliations have progressively larger per capita footprints, forming a cluster of similar values. The English have the highest per capita footprint, statistically higher than that of any of their Amish neighbors. These results are consistent with the expectations of most outsiders about Amish resource use.

The fact that the per capita outcome is the one we expected, however, doesn't mean that we shouldn't consider other explanations. Table 3.12 also shows that for medium-sized families the per capita footprints of the five groups no longer differ, and in all the groups the individual carbon footprint has increased. The Swartzentruber advantage is erased, as is English profligacy. This suggests that other factors may be involved. One possibility is that parents in large families are inherently frugal, or there may be a strong economy of scale that makes adding another person to a large family have less impact on the family's carbon imprint. In medium-sized families, there are more resources to go around.

In addition, carbon footprints are only partly the result of spending too much money. They're also generated by things that don't seem excessive. Carbon is the great equalizer of environmental impacts. So even a Swartzentruber family with twelve buggy horses and fourteen children can run up a big carbon bill. Even a rural English family that eats meat three times a day can have a significant dietary carbon footprint. If an Andy Weaver family relishes keeping its home very warm all winter, the high heating bills may offset English driving habits. In other ways, such as in consumption of goods and services from the larger economy, many Amish households are not very different from their rural English counterparts. The Holmes County settlement is one of the wealthiest and most progressive Amish communities in the United States. It is not clear to what degree our results can be generalized to other Amish settlements. Nonetheless, the degree of overlap in household economy that we found raises some important questions about the environmental impacts that any home exerts on global resources.

Whence the Future?

The Amish as a people are not frozen in time, and there is no single Amish life program. Rather, the Amish constantly adapt to new situations and to the world around them. To an outsider, the superficial similarities among Amish groups hide a great deal of variability both between affiliations and among individual families. Amish populations everywhere are growing rapidly. Newly married couples no longer predominantly farm. As the Amish are drawn into more frequent and more prolonged contact with the larger society, they are confronted with social, economic, and moral pressures that blur the boundaries between their traditional culture and the wider world.

Rapidly growing Amish settlements face increasing competition from non-Amish populations for land and for markets. Over the past few decades, larger settlements have achieved a historically unprecedented level of affluence and degree of economic stratification, creating immense challenges for Amish ways of life. While some groups, like the Swartzentruber Amish, have managed to retain their frugality and moderation, their Spartan lifeways are not embraced by higher Amish groups. But as our data show, the Swartzentruber Amish are the only Amish group whose behaviors usually conform to non-Amish expectations of low resource use for this simple life. In their limits on travel and mobility, their rejection of most fossil fuels, their self-sufficient diets, and their minimal material consumption, the Swartzentruber Amish could be identified as living with a light ecological footprint.

The more progressive Amish, however, are increasingly difficult to distinguish from their English neighbors. In their material consumption, the New Order Amish are not statistically different from the rural English around them. The Old Order Amish are not far behind. The Amish make heavy demands on fossil fuels in heating, and they increasingly use modes of transportation that depend on modern technology. To succeed in the entrepreneurial marketplace, Amish farmers and businessmen must engage the larger economy and adopt similar attitudes and marketing strategies. This creates exposure to lifestyles of comfort, opportunity, and self-realization that mark the English way of life. These lifestyles are notoriously resource dependent, and only with considerable attentiveness and self-denial can they be light on the earth. New types of businesses and lifestyles that extract or consume natural resources may also reshape the relationship of the Amish to their foundational identity as caretakers of creation. This is especially true of new forms of agriculture, which have seeded lively debates among the Amish themselves about what it means to be good stewards of the land.

II *Working with Nature*

CHAPTER FOUR

 The Transformation of Amish Agriculture

Few images evoke a greater sense of the Amish connection with nature than the bucolic photos of agricultural landscapes that frequent calendars, postcards, and other tourist merchandise at Amish settlements across the country. These depictions rely heavily on a pastoral motif that, according to Valerie Weaver-Zercher, "creates rural life as Edenic and situates urban life as diametrically opposed to agrarian innocence, purity, and charm."[1] Such images appeal to urban residents because they create nostalgia for an aesthetically pleasing, simpler time. This "sentimental pastoralism," as the historian Leo Marx terms it, rarely touches on the messy realities of rural life, like crop failure, competition from agribusiness, loss of farmland, and grueling hard work. Instead, the images are "uncomplicated and apolitical" and focus primarily on "bucolic scenery, gentle animal husbandry, and a pleasing sense of connection to the seasons."[2]

For all their nostalgic appeal, however, agricultural landscapes are far from natural ones. Farming involves an intentional and prolonged disturbance of an existing ecosystem, and it can only be sustained by annual inputs. While modern agriculture has been astoundingly successful in feeding the planet's people and livestock, farming nevertheless interrupts ecological processes and contributes to watershed degradation, loss of biodiversity, soil erosion, and other negative outcomes. The midwestern United States, where nearly two-thirds of the Amish reside, has "one of the most productive yet highly modified landscapes on Earth."[3] In the decades since World War II, intensive farming of a small number of annual crops, especially corn and soybeans, has led to dramatic simplification and homogenization of the landscape. The industrial model of agriculture that has taken root during this period is predicated on the use of pesticides, artificial fertilizers, and ever more sophisticated machinery in the fields and in the barn. The mechanization of American agriculture and its focus on economies of scale have created tremendous challenges for small farmers. But remaining competitive has been particularly challenging for the Amish because of their limitations on technology use and their historical preference for small-scale, diversified farms rather than large farms growing only cash crops.

Until almost the mid-twentieth century, Amish farms were very similar to those of their English neighbors, small family operations where people worked together as an economic unit.[4] Amish resistance to using tractors in the field in the 1930s and 1940s began to separate them from mainstream agricultural production, but the "big squeeze" came between 1950 and 2000, when land prices surged and farm equipment for the field and the barn became much more sophisticated.[5] Over the same period, the Amish population quadrupled, resulting in too many people looking for scarce and high-priced farmland. Though Amish groups responded in different ways, modifications to the restrictions on farming equipment in the Ordnung proved to be slow and difficult, lagging behind fast-moving changes in the broader economics of agriculture. In the face of the growing mechanization and corporate scale of agriculture, many Amish men sought out nonfarm work in factories and in small businesses. By the early twenty-first century, fewer than 20 percent of the heads of household in most settlements still farmed full time.[6]

Not all Amish farmers relegated their plows to the auction block, however. Those continuing to farm have found ways to update the traditional mainstays of animal agriculture and row-crop farming, and they have creatively pursued new agricultural niches suited to their lifestyle. Poultry barns that can hold tens of thousands of birds have become commonplace in many settlements. Some Amish farmers have capitalized on the artisanal food movement and turned to the high-end niche markets of certified organic dairy and produce. Amish farming has never constituted a singular entity, but its contemporary forms are more diverse than ever.

Statistically, the Amish who have remained farmers register hardly a blip on the radar of agricultural productivity nationwide. Yet Amish farmers remain a distinctive presence in localized agricultural landscapes because of their use of horse-drawn equipment and the small scale of their farms. This chapter examines how the Amish model of farming has held up against the rise of industrial agriculture and how Amish farming choices weigh ecological concerns against economic and religious ones. We ask why most Amish farmers adopted chemical-intensive agriculture, how an Amish organic movement arose, and how farmers navigate new governmental standards for food security and new consumer preferences for healthy food. We also explore Amish responses to genetically modified (GM) seeds and assess the debate within their community over genetically modified organisms (GMOs). As we will see, the diversity of their farming choices and their responses to regulation add complexity to the perceptions by outsiders of the ecologically minded Amish.

The Social and Ecological Case for Small-Scale Farming with Horses

Writing on his acclaimed blog *The Contrary Farmer*, the late agrarian writer and farmer Gene Logsdon titled one post "Did the Amish Get It Right After All?" His attention had been grabbed by an article in a Lancaster, Pennsylvania, newspaper in the wake of the 2008 financial meltdown: Hometown Heritage Bank was having its best year ever, doing $100 million in business and not having lost a penny in twenty years. Why? Because 95 percent of its loans were to Amish farmers. Logsdon reflected, "What struck me the most was the fact that these farmers are buying farm land that can cost them ten thousand dollars per acre or sometimes more, and paying for it with horse farming. And because of their religion, the Amish do not accept farm subsidies that keep many 'modern' farms 'profitable.' "

Logsdon's observation that horse farming can still be profitable is borne out both by research and by the firsthand testimonials of Amish farmers. A 1989 study by agricultural scientists at Ohio State University concluded that "in general, Amish farmers are not suffering the financial difficulties which currently are driving so many non-Amish farmers off their farms."[7] More recently, a new generation of Amish farmers is finding that a modest investment in the right combination of product, technology, and niche marketing makes solid financial sense compared with hundreds of thousands of dollars of debt incurred buying the latest high-tech equipment. Framed in this way, Amish agriculture stands out as one of the only examples of traditional, small-scale farming that remains economically sound in the temperate zone of the world. Over the past half century, the Amish capacity to keep small farms viable has preserved millions of acres of farmland that might otherwise have been given over to industrial development or urban sprawl.

Economically sound horse farming does not automatically equate with ecological farming, however. The term "sustainable agriculture" has become a catchphrase in recent years, but the concept is so broad that it hides as much as it reveals. The agrochemical giant Monsanto, for example, calls itself a "sustainable agriculture company" dedicated to creating a world in which farmers use fewer resources while growing more food.[8] Monsanto's definition emphasizes production and efficiency and retains a central role for bioengineered crops. Most supporters of sustainable agriculture, however, position themselves as an alternative to the industrial model in at least two important ways. They emphasize the embeddedness of farming in a sociocultural context that uplifts community, promotes the values of moderation, frugality, neighborliness, and equity, and is committed to locally grown, healthy food. In addition, they see farming as sustainable when it aims to disrupt natural processes as minimally as possible, thus affirming Aldo Leopold's view that "a good farm must be one where the native flora and fauna have lost acreage without losing their existence."[9] Amish farming embodies this vision of a socially stable, ecologically based agriculture in some ways but not others.

The social benefits of Amish-style, small-scale horse farming have been extensively documented.[10] The limitation that farming with horses places on the number of acres that can be plowed "keeps a farm to a manageable human scale," reflected one Old Order farmer. He mentioned frugality, the work ethic, and the cohesion of the family unit as benefits of keeping farms to "a realistic limit." Importantly, farming with horses keeps husbands close to home, so that the wife and mother is not the sole disciplinarian and coordinator of children's chores and after-school activities. While formal labor exchanges such as threshing rings have all but disappeared from Amish communities, farmers still rely on one another for advice and mutual assistance in many tasks. One study in Iowa discovered that the limits Amish farmers place on growth contributes to social stability.[11] In the "bigger is better" logic of agribusiness, horse farming is an anachronism, but this overlooks the ways small-scale farming fosters rich social ties and nourishes community relations.

In addition to social benefits, certain types of horse farming have ecological benefits. Unsurprisingly, fossil-fuel use has been shown to be relatively modest on Amish farms, even though most groups use gasoline-powered engines on horse-drawn machinery.[12] Where hedgerows, fencerows, and buffers along waterways are kept intact, Amish farms maintain local biodiversity. Long before there were federal farm programs and agricultural-extension agents, Amish farmers developed a reputation for taking depleted soils and building up their fertility, mainly through crop rotation and manure application.[13] The combination of using manure as fertilizer and the general practice of "a four-year rotation of hay, corn, oats, wheat or barley" produces "a friendly environment for soil life," noted one veteran Amish farmer.[14] John Kempf, an Amish farmer in Middlefield, Ohio, has received mainstream media coverage as the unlikely CEO of a consulting firm that promotes "regenerative agriculture."[15] Kempf argues that farmers need to actively rejuvenate dead soils by adding minerals and nutrients to ensure that crops grow well. While this holistic approach to soil fertility has its critics—one soil and water conservation agent we talked with scoffed at some soil amendments as "magical products"—Amish farmers such as Kempf have become adept practitioners of nontraditional soil science.

Farming with horses also allows more intimate contact with the weather, the soil, and the natural world than does tractor farming. "It teaches you a lot about nature and life, about the earth itself," reflected an Old Order farmer. The director of a small-farm-focused nonprofit organization told us, "It's so different from the farmer who is ten feet off the ground in a tractor and doesn't know what's going on in the fields because he gets a crop consultant to tell him which seeds to buy and when to fertilize." The economics of small-scale farming are tight, so Amish farm families are often attentive to other farm resources, like maple-syrup production, that can generate supplemental income from their land.

Beekeeping is an ecologically beneficial sideline business pursued by many Amish farmers that leads to a keen appreciation of one's natural surroundings. The industriousness and division of labor exhibited by honeybees exemplify qualities that are valued in Amish culture. Beyond their economic motivation, Amish managers of bee yards routinely become careful observers of nature. One Ohio beekeeper said that his bees made him more attentive to what plants were blooming. "In the spring, it's like, 'Well, coltsfoot, wow, they're out already.' Yeah, you're very observant to the flowers, and I like that." Amish apiarists know which nearby farms are organic and which are not, and they are aware of new threats to their hives, whether from pests like varroa mites or the malaise known as colony collapse disorder.[16] An Illinois beekeeper who had experienced hive loss was taking a proactive approach to protecting his hives by collecting feral bee swarms that gathered on a stump on his farm.[17] His goal was to find "a survival colony," on the premise that if a hive was healthy enough to produce a swarm, those bees might resist other threats that challenge beekeepers.

Keeping small farms alive and showcasing farming as an attractive occupation for young Amish families have been central goals of celebratory annual Family Farm Day events in Ohio and Pennsylvania. Open to the public but organized by a mostly Amish committee and hosted outdoors on an Amish farm, the weekend event in Holmes County includes a motivational keynote speaker and hands-on demonstrations of sheepherding, butchering, and horse training. In Lancaster County, the event addresses farming and health and attempts "to provide an educational basis that promotes down to earth ways to bring forth nutrient dense food for homesteaders and farmers alike."[18] The main focus of both events, however, has been on children and young adults. "We were finding that young couples, lunchbox Amish, would move into new homes and didn't know how to garden because they were two or three generations away from the farm," explained one of the original organizers of Family Farm Field Day in Ohio.

The fact that farming must now be consciously promoted within the Amish community as a viable occupation for young families is a testimony to how powerful the lure of off-farm employment has become. In Amish circles, husbands, not wives, generally make the decision to farm. One woodworker acknowledged that his girls did not have the same mind-set as that of girls raised on a farm: "Would our daughters attract boys who wanted to farm? I don't know. When I was a kid, everybody in our school was a farmer. If you didn't farm, it was like, 'What are you doing?' But now, among our Amish, there are kids who see someone doing hand-milking and it's like, 'What are *you* doing?' " In spite of these generational changes, promoters of small-scale farming argue that their size allows for a certain nimbleness. "Once you have the infrastructure to support a thousand cows, it's hard to change," said one Pennsylvania farmer. "But with our small farms, we have a huge advantage." Over the past few decades, however, that nimbleness has paradoxically taken Amish farming in several different directions that simultaneously enhance and threaten the model of ecologically sustainable farming.

Chemical-Intensive Agriculture, Amish Style

Many outside observers assume that all Amish farmers are organic farmers, but the reality is far more complex. Though a small, if robust, organic movement has arisen in Amish circles in the past twenty years, the majority of Amish livestock, grain, and produce farmers, like their non-Amish counterparts, use artificial fertilizers and pesticides to supplement manure application and crop rotation. Why is it that the Amish church leadership resisted the tractor and other self-propelled farm equipment in the mid-1900s but allows chemical-intensive agriculture today? Part of the answer is that the Amish have always been on the lookout for methods that would increase production but have minimal negative impacts on family or community life. The application of chemicals did not raise any red flags for the church leadership because it came without the publicly visible indicators they associated with modern agriculture, such as electrification or self-propelled machinery with rubber tires. A cultural preference for neatness also played a role in the acceptance of chemicals such as 2,4-D and Heptachlor, according to many farmers. The Amish are "very Germanic, they want straight, clean fields," one Old Order farmer pointed out. "You really see it in Lancaster [Pennsylvania], everything is perfect." An Ohio dairy farmer recalled that "one year a guy used Atrazine and he had clean corn, and the next year everybody switched." The ever-cautious Swartzentruber Amish even joined the tide, allowing sprayers in the field in spite of holding the line on virtually every other type of new farming equipment.[19]

Another important factor contributing to Amish adoption of chemicals was that agricultural-extension agents vigorously promoted the new techniques. Accord-

Spraying pesticides and herbicides on crops is a widespread practice among Amish of all affiliations, who use the chemicals to achieve higher yields. Photo by Doyle Yoder

ing to one Ohio organic farmer, farm magazines in the late 1940s and 1950s "were absolutely in awe of herbicides and pesticides" and touted them as "the answer to mankind's problems." The Amish had seen the benefits of penicillin and antibiotics, he noted, and had no reason to question the judgment of experts. In this case, he concluded, "we trusted science. We were never, ever warned of any dangers." Finally, the dramatic changes in American agricultural productivity between 1950 and 1975, based on new hybrid seeds, technological innovations, and a rapid increase in the use of pesticides and fertilizers, came just as Amish farmers were struggling to survive. The shift to off-farm work was growing, and small-scale farmers were under great pressure to show that they could make the economics work in their favor. "I know there are people who think that if they buy Amish food it's the purest thing they can get," reflected one Old Order man who grew up on a farm in Michigan. "But the Amish educated themselves on how to get the best crop, and it actually involved a lot of chemicals."

In the decades since the adoption of chemical-intensive agriculture, technology use on Amish dairy farms has also changed, but in an increasingly diverse manner.

The Swartzentruber Amish and other conservative groups have remained in farming to a greater extent than other groups and have accepted the economic consequences of resisting new technologies. Many continue to milk a small number of cows by hand and to find ways to ensure a market for their can milk.[20] Dairy farmers in conservative groups frequently supplement their farm income with sideline businesses, including the sale of strawberries, maple syrup, firewood, eggs, and woven baskets. Many still slaughter beef cattle, hogs, and chickens for family use. The Swartzentruber Amish, forbidden to work within the city limits or for non-Amish employers, have also maintained generalized labor exchanges with their neighbors, and Swartzentruber men participate in daily household activities to a greater extent than men in other groups.[21]

At the other end of the spectrum, dairy farmers in technologically progressive Amish groups have adopted a number of crucial technologies that allowed them to stay competitive in the Grade A milk market, such as vacuum milking machines, bulk cooling tanks, electric fences, round balers, and skid-steer loaders. In the Lancaster settlement, many dairy farmers have expanded their herds from forty to eighty cows and rely on seed consultants and agronomists for advice on soil management and the application of liquid manure, chemical fertilizers, and pesticides on their grainfields.[22] "A lot of the Amish around here really are part of the local agricultural community and tend to follow the big trends," observed a Pennsylvania State University extension agent. Many subscribe to agricultural periodicals and even belong to the American Farm Bureau Federation, a national lobbying organization that is often seen as a voice for Big Agriculture.[23] Nor are these Amish lacking in modern farm implements. Amish entrepreneurs have designed new farm equipment that can be powered by gasoline engines and pulled by horses. The notion that most Amish farmers use outdated technology is "a false impression," said one Old Order farmer, "because even though we farm with horses, we still have modern technology right at our fingertips."[24]

One relatively new face of Amish farming that began in the 1970s involves raising poultry for eggs or meat. For Amish farmers, the appeal of raising poultry is that it can be done on relatively small acreage and within the limits of acceptable technology available to more liberal affiliations. Though the term "Amish chicken" usually evokes images of chickens ranging freely on green pastures on a small family farm, the reality is somewhat different. Most Amish poultry barns are long, enclosed buildings that can hold twenty to thirty thousand layers or broilers; in the case of the latter, they range freely for the six weeks of their lifespan only within the confines of the structure's four walls. Lighting, heating, and fans are run by generators, and the feed is a mixture of GM corn and soybeans. Amish poultry farmers typically work under contract with a company such as Gerber's Poultry or Miller's Amish Country Chicken, which provides chicks, feed, and veterinary care, as well as slaughtering and packaging services, and the use of antibiotics depends greatly on corporate policy. In terms of how they raise chickens, Amish and non-Amish poultry farmers are hardly distinguishable from each other, yet poultry corporations, supermarkets, and upscale restaurants continue to use the label "Amish chicken" to capitalize on the aura of naturalness associated with the Amish community.[25]

In light of changes over the past few decades, it seems necessary to rethink the label "small-scale" as it applies to some Amish farms. Though almost all Amish farms remain relatively small in terms of land, some are becoming similar to industrial farms in the increasing technological efficiency of their dairy operations, their use of chemical fertilizers and pesticides on crops, and the confined feeding approach they use in raising poultry (and hogs). Some Amish farmers are even growing the same crop on the same field for multiple years, a simplification that the food critic Michael Pollan says "is poorly fitted to the way nature seems to work."[26] Hedgerows and buffers are sometimes plowed under as farmers try to maximize the productivity of their acreage. Moreover, the impact of runoff from Amish farms on water quality has attracted the attention of regulatory agencies, a story we explore in chapter 10. In order to remain competitive, many Amish dairy and poultry farmers have adopted some aspects of the factory-farm model. Drawing the contrast too sharply between "sustainable" Amish and "unsustainable" non-Amish farms runs the risk of minimizing or overlooking these changes in Amish farming methods that are less ecologically sound.

At the same time, Amish poultry and dairy operations do not approximate the size and technological sophistication of industrial-scale concentrated animal feeding operations, which may house more than two thousand head of cattle. Nor does the acreage Amish farmers plant in crops come close to the size of large non-Amish corn and soybean operations. The limits Amish farmers have thus far placed on growth and expansion affirm the conclusions of the Dutch scholar Martine Vonk, who argues that "while not entirely sustainable in the ecological sense," Amish farming "certainly has aspects that lead to a low environmental impact."[27] One need only drive through Amish farm country and compare it with the vast expanses of sterile fields planted in corn and soybeans throughout the Midwest to be reminded that, as Wendell Berry says, "in the Amish country, there is a great deal more life: more natural life, more agricultural life, more human life."[28] The attempt to nurture that life through certified organic farming by a small segment of the Amish population deserves a closer look because it represents a distinct break from the midcentury adoption by the Amish of a chemical-intensive path.

The Rise of the Amish Organic Movement

The origins of the Amish organic movement in the 1990s must be seen in the context of the ongoing struggle to find economically viable products that fit with the scale and values underpinning the Amish farming model.[29] By the late 1980s some influential opinion leaders in the Amish community had begun to question the long-term effects of chemical-intensive agriculture. In 1990, a three-part series in *Family Life* entitled "Poisoning the Earth" sharply criticized the use of chemicals on Amish farms as a "radical departure" from traditional ways of farming.[30] At roughly the same time, the economics of producing certified organic milk began to shift enough to justify refraining from using synthetic fertilizers, herbicides, and pesticides on fields for the three years required to transition them to organic farming.[31] For example, Wisconsin-based Organic Valley, a milk cooperative with strong farmer input, began offering a guaranteed market for certified organic milk to dairy farmers in several midwestern states. Among the first Amish pioneers were influential Ohio farmers like David Kline and the late Rob Schlabach, who could never quite reconcile application of synthetic fertilizers and pesticides with their view of stewardship of the land. "They were willing to take the leap," remembered one New Order man. "They're the ones who said to their fellow farmers, 'Here's organic milk, here's what we can get, and we actually believe this is a better way of farming.' "

For these early Amish converts, organic dairy farming offered a chance to reclaim an identity that had been lost in the rush to embrace chemicals.[32] As one farmer put it, "Organic rings a bell with us. We invented that almost. That's who we are." Most Amish organic farmers we spoke with, however, acknowledged that economic incentives were crucial in their decision to change. In this respect, the multiple motivations of profit, environmental stewardship, and personal health moved both Amish and non-Amish farmers toward organic grains.[33] The move to wider acceptance of organic methods in the Amish community has been slow, however, because farmers who adopted chemical-intensive farming have come to see their way as "traditional." The handful of Amish farmers who resisted chemicals initially had few reliable markets for their products, and their fields were widely viewed as unproductive. "Our parents valued farms and homesteads where everything was very tidy and clean. But organic farms in that day had weeds and were scruffy," remembered a man who grew up on a farm in Michigan. Organic farming was also

Produce auctions and farmers' markets around the country provide the Amish with an ideal outlet for their vegetables, baked goods, and other items. Photo by Doyle Yoder

associated with the hippie movement: an oft-heard joke was that the three-year transition to organic was just long enough for a farmer to grow a ponytail. Even though the image of organic farming has become much more positive over the past two decades, many Amish still see organic farming as requiring too much work.

Nevertheless, a small but growing minority of Amish farmers have hitched their wagons to the farm-to-table movement and believe the future belongs in niche marketing to well-heeled consumers who are willing to pay a premium for organic products. Organic farming among the Amish has thus followed the national trend, which saw sales of organic livestock and poultry products double between 2008 and 2014.[34] The prevalence of organic farming among the Amish varies somewhat by settlement and region, though. In Lancaster, Pennsylvania, an agricultural-extension officer who works closely with the Amish estimated that 5–10 percent of farms were certified organic, while an agricultural consultant in Holmes County, Ohio, put the figure at 10–15 percent. These figures suggest that more than 85 percent of Amish farmers across North America still use synthetic fertilizers and pesticides to some degree. Yet Amish organic farms represent a higher percentage of the national total of organic farms than one would expect given that the Amish make up only a fraction of 1 percent of the US population as a whole.[35]

Milk and eggs make up most of the organic market, but some Amish farmers have discovered another profitable product that requires relatively small parcels of land and benefits from family labor: organic vegetables. The economic viability of produce farming in general received a huge boost in 1985, when Amish growers first organized an auction house in Lancaster, Pennsylvania, where wholesale buyers such as grocery stores and restaurants could bid on Amish-grown fresh vegetables, fruits, herbs, and flowers. Driven by consumer demand for wholesome, locally grown food, the model caught on like wildfire, and by 2017 more than sixty Amish-organized produce auctions had appeared across the Midwest.[36] Most of the produce sold at these auctions is grown conventionally, but organic produce occupies an increasing share. A closer look at one Ohio-based organic produce cooperative, Green Field Farms, illustrates how Amish growers balance market forces, cultural and religious traditions, and ecological concerns.

Inside an Amish Organic Produce Cooperative: Green Field Farms

When Amish community leaders in the Holmes County settlement in Ohio came together in 2003 to discuss the idea for a produce-farming cooperative, the main focus was on how to make farming attractive again and stem the migration of Amish men to off-farm work. "It's about preserving a way of life," said one board member. "Farming keeps us from getting too close to the edge." The founding group concluded that the best way to make small-scale farming viable was to create more value on fewer acres and charge a premium for their products.[37] According to one long-term employee, "Organic wasn't even considered in the planning stages, but after we wrote up the specs, they were almost at the level of certified organic. So we said, 'We might as well do it.'" Thus was born Green Field Farms (GFF), a cooperative "dedicated to sustainable agriculture for the horse farming communities of Amish and conservative Mennonite people."[38]

The challenges to realizing this vision were considerable, however. At a minimum, GFF needed to find markets for its growers in a very competitive economic environment while still respecting "the moral boundaries of Amish culture[, which] severely restrict some types of advertising."[39] In addition to prohibitions on television and radio advertising, the use of the Amish name to promote a product has long been frowned on as "using our religion to sell something." Without advertising, however, no consumer can tell the difference between, say, a cucumber grown

The Green Field Farms Seal of Approval represents a subtle marketing strategy that calls attention to the Amish origins of the cooperative's organic foods without using the Amish name. Courtesy of Green Field Farms

by an Amish farmer and one grown by a non-Amish farmer. One way the GFF leadership solved this problem was to cultivate direct relationships with buyers by inviting them to orientation days that included field walks and lunch in an Amish home. As the cooperative's director stated, "Our goal is to make it memorable for them. If you serve them a meal in an Amish home, that's the highlight for them." Another effective marketing strategy has been the development of a logo, dubbed the GFF Seal of Approval, which shows a team of horses plowing a field as the sun sets, highlighted by the words "Certified Organic." Below this image is a silhouetted horse and buggy with the word "certified" in capital letters. The cleverly designed label harnesses what the anthropologists John and Jean Comaroff have called "the identity economy."[40] It subtly calls attention to the Amish origin of the product, with all its connotations of naturalness and quality, while skirting cultural disapproval by not directly using the Amish name. As the GFF director of operations confided, "From a marketing standpoint, we all understand that the Amish name sells."

Another challenge was specific to growing certified organic produce: how to replenish soil nutrients without synthetic fertilizers. Small-scale dairy farmers grow grains and forage that feed their cows, whose manure replenishes nutrients in the fields in a self-reinforcing loop. Because produce growers sell their crop off the farm, they break this nutrient loop, so they are forced to employ natural fertilizers. GFF strongly advocates the use of cover crops, which can be plowed back into the soil, as one solution to this problem. Since green manures usually cannot satisfy all nutrient requirements, however, a second solution has been to construct their own soil-amendment facility, where bags of raw minerals can be custom blended in an industrial-scale mixer. GFF sends soil samples from growers' farms to a local lab for analysis, and then a self-trained GFF agronomist determines the exact nutrients the farmer needs, rather than relying on preblended commercial formulas. The philosophy behind the soil-amendment program, which features calcium as the king of macronutrients rather than the conventional big three—nitrogen, potassium, and phosphorous—and focuses on trace minerals such as sulfur, boron, and zinc, has proved popular with GFF's growers. Many farmers report that soil amendments reduce weeds and pests and produce higher-quality vegetables.

Finally, GFF had to navigate a plethora of government regulations around organic certification and food safety, which posed special challenges for Swartzentruber Amish growers, with limited technology for safely washing and transporting vegetables. Organic food is usually contrasted with modern, chemical-intensive food production, but insofar as "grades and standards are part of the moral economy of the modern world," the adoption of certified organic farming takes the Amish into unfamiliar territory.[41] It sets new norms for behavior and seeks to create uniformity in products, workers, and markets based on standards imposed by a distant federal or state government and verified by outside inspectors. In 2011 the federal Food Safety Modernization Act laid out a sweeping set of new protocols that included "enhanced traceability systems" to identify points in a product's life where a pathogen could potentially enter.[42]

Green Field Farm's response to these federal guidelines has been instructive. To assist with recordkeeping, it created its own operational checklist for farmers, which one of Ohio's chief certifying agencies, the Ohio Ecological Food and Farm Association, adapted for its own use. GFF also arranged to pick up all produce at the farm, thus allowing growers to maintain social distance from consumers, in exchange for farmers' upgrading their systems for washing vegetables and taking more stringent steps to ensure that manure is kept out of the food chain. The co-op has persuaded farmers to go along, not through a newfound sympathy for government regulations, but by appealing to the power of market forces. As the keynote speaker at an annual GFF growers' meeting put it, "Lots of people can get upset [about government regulations], but when the dust settles, someone else will make a decision about food safety—the consumer!"

The success of GFF in overcoming these and other challenges has resulted in growth that far exceeded expectations. Fresh produce, eggs, and even kale chips with the GFF label are now sold at mainstream and upscale retail outlets across the

Midwest, including Kroger and Whole Foods. In a relatively short span of time, affluent consumers throughout the region have gained access to organic produce grown by more than one hundred Amish farmers, roughly half of them from the most conservative affiliations. For their part, farmers can net anywhere from one thousand to five thousand dollars per acre, far more than they would receive for conventional commodity crops. Moreover, they share in the overall profit of the co-op. "That's the clincher," said the director, adding that the co-op usually gives back 4–5 percent. "Why shouldn't they be happy? They're making a living at home with their children instead of dad packing a lunch and going off to work each day."

As the Amish growers and board members have interacted with environmentally conscious consumers, many Amish have become more convinced of the health and ecological benefits of organic foods. Rather than embracing organic as a secular philosophy, though, Amish members of GFF see it as a form of stewardship of God's creation. Citing a study showing that organic foods were nutritionally superior, one columnist wrote in the GFF quarterly newsletter, "For farmers, I believe it shows dedication to doing things the way the 'Good Lord' has intended it to be done."[43] It is also telling that even while meeting the highest standard of ecologically sound agriculture, GFF still frames its effort largely in terms of the economic benefits to the Amish community rather than invoking an explicitly environmental perspective. One long-term employee told us the co-op defines sustainability by asking a single question for each of its growers: "Is this family living on the farm, able to make payments, and hopefully has something better when they pass it on to their children?" Nevertheless, the case of Green Field Farms illustrates that when ecological agriculture can be made economically profitable, some Amish farmers will find a way to get behind it.

God Move Over? The Amish and the GMO Debate

On September 10, 2011, nearly two hundred individuals, roughly half of them Amish, gathered in Dalton, Ohio, for an event sponsored by Right to Know, a nonprofit organization designed "to educate the public about Non-GMO."[44] In addition to a panel discussion on the dangers of GMOs, attendees heard two keynote speakers deliver passionate attacks on the biotechnology industry and on government policies that support unlabeled, GM food. Audience members learned about the difference between open-pollinated, hybrid, and genetically engineered seeds.[45] And they discovered that while USDA certified organic food is almost always non-GMO, the non-GMO label does not necessarily imply an organic product.[46] In between sessions, attendees mingled with one another and visited information booths promoting natural foods and organic products. "Taking genes from one species and putting it into another is new," commented one. "They think they're God," chimed in another, referring to Monsanto's practice of patenting GM seeds. "All the corn around here is Round-Up Ready," said a third who farmed organically. "Just look at the signs in the fields."

Of course, those signs that indicate the planting of GM crops stand in the fields of most English *and* Amish farmers. In the late 1990s, many Amish farmers joined the larger agricultural community in adopting Bt corn, whose seeds have been genetically modified to synthesize an insecticide compound that kills the European corn borer caterpillar. The Amish were especially drawn to Bt corn because without the corn borer the stalks stay intact (rather than breaking in half) when they go through horse-drawn binders, effectively cutting their harvest time in half. Most corn planted by Amish and English farmers has also been genetically modified to resist Round-Up, the brand name for the herbicide glyphosate, so this herbicide can be sprayed on surrounding weeds without harming the crops themselves. Amish farmers who use GM seeds cite increased yields as the main reason for using them. As a Lancaster, Pennsylvania, farmer told a BBC reporter: "We have to keep farming in a way that is both economical and practical. As a church group, we are not opposed to GMOs. It's just a tool that we're using in the same way that we use pesticides. We're getting very high yields and great returns per acre."[47] An agricultural consultant we spoke with concurred that in general the Amish tend to be open to GM, or what he called "traited," seeds. "They see the need to keep in line with the English in order to compete."

Important nuances exist around the Amish use of GM crops, however. Though soybeans and corn together account for approximately 90 percent of the GM crops in the United States, Amish farmers devote very few acres to soybeans because they are virtually impossible to harvest with horse-drawn machinery. In addition, the most conservative groups often do not use GM seeds, primarily because of the cost and a perception that they may cause health problems. A Swartzentruber farmer told us that deer wouldn't touch a neighbor's field of Round-Up Ready corn but trampled all over his corn while eating it. "What does that tell you?" he asked. He saw no contradiction between avoiding Round-Up Ready corn, however, and being licensed to spray the herbicide Atrazine on his own fields. But Amish farmers who use GM seeds are not completely deaf to ecological concerns. Some cite the decreased necessity of spraying broad-spectrum insecticides as a tangible environmental benefit of Bt corn. Pressed on how he would reply to someone who says GMOs will lead to an environmental disaster, the Lancaster farmer interviewed by the BBC concluded, "I would say they're misinformed. They don't know what they're talking about."[48]

Undeterred, Amish critics of GMOs such as the ones who attended the Right to Know conference promote alternative perspectives similar to those voiced by non-Amish opponents of GMOs. For example, they point to the rise of "superweeds," which are resistant to Round-Up and result in ever-heavier annual spraying of herbicides that can alter microbial life in the soil. The Amish have been particularly receptive to arguments about the health dangers of GMOs. "I like organic because I think GMOs are causing a lot of diseases," commented an Amish man in Colorado. *Plain Interest*, a magazine widely read by more conservative Amish, has run a number of columns that raise questions about GMOs. A 2011 article in *Family Life* with the ominous title "Is There Death in the Pot?" argued that there is a connection between the use of hybrid and GMO seeds and an increase in mental and physical illnesses. "On my part," the author concludes, "I want to stop worshiping at the table of science and go back as nearly as possible to God's way. To me, that means open-pollinated seeds, no sprays, no chemicals."[49]

Whether or not one is sympathetic to the author's argument, his criticism of science as part of the problem ignores the fact that Amish critics of GMOs frequently cite scientific studies to make their case. Nor does it mention that hundreds of scientific studies have yet to show any negative health consequences for people and animals eating GM food over the past two decades.[50] In this sense, the GM debate in the Amish community is similar to the wider societal debate, in that bioengineering has come to symbolize either the promises or the perils of industrial agriculture, depending on one's point of view. As the anthropologist Glenn Davis Stone points out, both sides approach GMOs as if "they are a single entity up for approval or rejection" even though each "transformation event" is unique and should be evaluated on its own.[51]

Overall, though, the Amish debate over GMOs has been fairly muted. Skeptics have by and large preferred to direct their critiques to sympathetic audiences. "I'm not going to judge them," said one seed dealer, who had no qualms about selling hybrid and traited seeds alongside his line of organic seeds. For their part, Amish farmers who use GM seeds have kept a low profile, perhaps sensing that bioengineering runs counter to the public image of the environmentally conscious Amish or that it gives humans a role that was intended for God and therefore pushes the limits of "acceptable genes."[52] Instead, like their non-Amish counterparts, most Amish farmers seem comfortable with the dominant societal narrative that sees plant manipulation, including genetic modification, as progress.

Flexible Farmers

The Amish have long been known for diversified farming, whereby a family avoids specializing in one commodity and instead follows the tried-and-true model of keeping a few dozen dairy cattle, growing multiyear rotations of grain crops, using manure as fertilizer in the fields, raising a few hogs, chickens, and beef cattle, and planting a vegetable garden. As external economic pressures have intensified and as the Amish world itself has become more diverse, however, farmers positioned differently across affiliations and settlements have pragmatically pursued new agricultural opportunities, each having its own set of ecological ramifications. Assessing the relationship between sustainability and this diverse terrain of agriculture, Kraybill, Johnson-Weiner, and Nolt acknowledge that "Amish farm technology and practices are not driven by environmental concerns," but they argue that "Amish values and way of life tend to mitigate environmental harm."[53] An Old Order Amish man in the Arthur, Illinois, settlement put it a bit differently: "When it comes to agriculture, we've always been utilitarian. We substitute flexibility for sustainability."

At one end of this spectrum of flexibility, the small but growing number of Amish organic and non-GMO farmers represent a case where consumer perceptions of the Amish as environmentally conscious closely align with what are widely regarded as ecologically sound practices on their farms. The premiums paid for organic products provide an economic incentive for Amish farmers to meet the high bar set by the USDA for certified organic products. Moreover, adopting the organic brand brings tangible benefits to their community by preserving a way of life tied to the farm. Amish organic farmers have proved to be astute observers of consumer discourse around sustainability, and they have learned how to capitalize on the public perception that the Amish farm "naturally."

When ecological and economic incentives conflict, however, a different story unfolds. Like other agricultural operations, Amish farms have come under increased scrutiny in recent years for manure and nutrient runoff that contributes to the pollution of regional watersheds. Similarly, participation by the Amish in the timber and furniture industries, to which we turn next, has the potential to impact forests outside their immediate communities. In these instances, Amish are asked to understand that the actions they take on their own land and in their own businesses may, in an incremental way, impact the health of larger ecosystems and thus the quality of life of Amish and non-Amish people whom they may never meet. Regulatory agencies sometimes require them to make economic sacrifices in order to restore those ecosystems and thereby contribute to the larger public good. Amish reactions to such challenges provide a useful barometer of how they view the balance of economic, social, and environmental priorities.

The Forest for the Trees

The Wood-Products Industry

Like farming, forests are part of the Amish DNA. Wood is an essential element of the rural lifestyle, and forests have long been part of the Amish environment. The rolling agricultural landscape of the eastern United States is peppered with woodlots, and the Amish make ready use of them. Even in western states, where most of the timbered land is publicly owned, the Amish travel to forests for wood, provisioning, or recreation. "An Amish man's dream farm," declared one Amish sawmill owner, would include "fifteen to twenty acres of forest, maybe a little creek or pond." Forests provide firewood for household use, a place to hunt, a site for collecting mushrooms, hickory nuts, berries, or maple sap, a playground, and a haven of solitude and beauty. The Amish value forests for many reasons, and those who engage in outdoor activities gain a deep familiarity with the forest environment. Like many Americans, the Amish often see forests as restorative, beautiful, and even inspiring.

Forests are especially valuable for making a living. Timber is an important commodity, and forests drive an entire production chain, from the tree trunks skidded from the woods to the grade lumber from the sawmill and ultimately the fine furniture and other products into which it can be transformed. Amish households and communities have always produced wood for their own homes or for local use, and they have a long history of craftsmanship in cabinetry and furniture-making. But as they seek new livelihoods off the farm, their engagement in the commercial forest products industry has grown dramatically.[1] According to one New Order lumber dealer, as many as half of Amish men in the Holmes County settlement are employed in the wood sector—in harvesting and milling lumber, in home construction, and in wholesale and retail sale of wood products.[2] Most Amish enterprises are small, with just a few employees, but a growing number of Amish-owned wood-products businesses employs a larger work force.[3] Their familiarity with wood and their strong work ethic make the Amish a natural fit for such positions.

Furniture is probably the best-known wood product made by the Amish. It is widely valued as the work of careful craftsmen who hand-fashion sturdy, beautiful, and useful items for an increasingly urban English population.[4] Amish involvement in other parts of the forest industry, however, has not been well documented. The primary forest sector, which is largely invisible to the consumer, involves harvesting wood from forests and converting it into lumber or other products, such as pulp, sawdust, or woodchips. Timber buying, logging, sawing, planing, and kiln drying are all primary-sector activities that have attracted enterprising Amish workers. From the products generated in the primary sector, workers in the secondary forest sector fabricate an astonishing variety of products. Capitalizing on skills that are part of their heritage, Amish businesses make everything from railroad ties and pallets to garden sheds and stakes to toys and birdhouses.

Because of the close association between Amish and forests, one might assume that careful forest husbandry lies behind many Amish wood products. An article titled "The Green Machine," published in an Ohio tourist magazine, makes exactly this assertion: "From raw timber out in the woods to the finished pieces of furniture displayed in showrooms, Amish furniture is built one step at a time—all with eco-friendliness in mind." In contrast to "some companies," the article continues, Amish loggers harvest trees using "responsible practices" and "without disturbing the environment's natural habitat."[5] In this particular version of Amish branding, the claims go well beyond the usual generalities about the wholesomeness and authenticity of Amish furniture, referring to specific practices and outcomes in the primary forest sector.

In this chapter, we explore how well the claims in "The Green Machine" match what we heard from foresters, sawmill owners, and woodworkers. We examine the supply chain for the Amish wood industry, from the standing tree to the dining-room table. Throughout, we ask how the wood-products industry reflects Amish views of nature and to what extent Amish who buy, cut, and saw timber follow ecologically sound practices. Not surprisingly, we discovered a spectrum of practices within the Amish wood industry but also a widely shared understanding that forests should be, first and foremost, for human use.

Buying Timber

A quarter-sawn red oak dining-room table in an Amish furniture showroom begins as a tree from twelve to forty inches in diameter, somewhere in the deciduous forest.[6] In the eastern United States, most available timber comes from private tracts of forest. In Ohio, for instance, private owners control 85 percent of the forested land.[7] Many Amish also own forested tracts, but the vast majority of private tracts are English owned. The wood supply chain therefore typically begins in a woodlot.

The first step toward that Amish dining table is an Amish or non-Amish timber buyer's visit to woodlots, where he estimates the value of the trees he believes should (or could) be cut from that site on the basis of the cutting regime he expects the loggers to use during harvest. After conferring with the sawmill owner, he makes an offer to the seller, who can decide to take the bid, modify the contract, or otherwise negotiate the terms of sale. Timber buyers may work on their own, bidding on standing timber and then selling the contract to a logger. More commonly, they work directly with one or several sawmills and evaluate a forest tract in terms of the kinds of logs their mills might want to acquire. A good timber buyer has extensive experience in evaluating standing timber, as well as connections within the primary logging sector that will help direct logs to the right places. Working with drivers, Amish timber contractors bid on sales over a sizeable geographic area. They advertise in print media that they "buy standing timber." In the Holmes County, Ohio, settlement, an Amish sawmill owner estimated that 70–80 percent of timber buyers within a hundred-mile radius of his mill were Amish. A non-Amish forester agreed with that estimate.

Timber can be valuable, depending on the species, size, and health of the tree. But trees grow slowly, and most landowners will cut their land only once or twice in their lifetime. Many know little about the logging process and about what their trees are worth. Because they must depend on advice from someone who knows the industry, they are vulnerable to dishonest timber buyers. Sometimes a buyer will knock on the door and offer a contract on the spot, pressuring the landowner to sign right away to keep the deal alive. One forestry professional told us: "There are a number of guys, they go to a landowner and offer, say, $25,000 for their timber, get 'em under contract, buy the trees, and then turn around and go sell it to somebody else, for usually a pretty big profit. That happens a lot. Most of these contracts that are written up by a guy like that are about three paragraphs. It's not going to have any detail in it. Once you sign, that will hold up in court no matter what." An Old Order sawmill owner in Ohio agreed. "The timber industry has gotten a bad rap," he said, "because a lot of guys are not honest." He added that timber buyers who offer a price for a forest and then broker it to a logging firm to do the cut are the worst. "These are the problem folks. They are basically dishonest. The opportunities are huge to make a killing on one sale. A lot of temptation here."

From an environmental standpoint, the main impacts of a logging operation on the physical environment have to do with what trees are cut and how. Amish timber buyers or logging contractors make offers for standing timber based on their assessment of what they can harvest, how much the trees will bring when sold up the supply chain, as well as the cost of the logging itself. Some landowners sell timber simply because they want the money, while others want to manage their forests for wildlife or to improve the value of their future timber harvests. The motives of both timber buyers and landowners determine how forests are cut.

In the eastern United States, there are three main models for timber harvesting, with different consequences in terms of profit and forest health. The first model, used by many Amish timber crews, is a "diameter cut." Loggers set a minimum tree diameter (12–16 inches) and then take every tree of every species whose diameter meets or exceeds that minimum. This approach is good for the buyer, because it maximizes the number of stems he takes from the woods. It may be good for the landowner, because it maximizes his earnings. It's also good for the logging operation, because it's easy to implement. But it is a very heavy cut. If the goal of good forestry is to retain an ecologically intact forest, this method fails the test. When all the larger trees are removed, most of the canopy disappears, resulting in too much light reaching the forest floor and too many seedlings competing to regrow. What remains is no longer a forest. In the near term, the site will become an overgrown, shrubby mass of brambles. Although the forest will return, it will take forty years or more for the site to support another timber cut.

A second model, which is hard on the forest for different reasons, is "high-grading," or "creaming." This cut is good for the Amish or English timber buyer, but not for the landowner. Loggers are tempted to high-grade when they suspect that landowners don't know the value of their trees. The loggers minimize their efforts and maximize their profits by taking only the largest, straightest trees of the most valuable species. While they don't take as many trees, the ones they cut are particularly valuable. The forest is left with trees that are of poor quality, diseased or damaged, and of species that are worth little but remain to compete with the more valuable seedlings. This method leaves the forest degraded aesthetically, ecologically, and economically. An Amish sawmill owner who deplores this practice told us: "In the past, the loggers came in and just took out the trees that they really wanted to take, the high quality genetics, and left the other stuff. . . . They leave all the beech, and the gum, and the trees that are not worth much, and a beech tree will crowd out the others until all you have left is beech." Furthermore, the landowner is likely to be substantially underpaid for his timber.

The third option is a "selective harvest," also known as a "management cut."[8] Both Amish and English loggers can do this, but they must have a superb understanding of forestry to do it well. In selective harvesting, the buyer evaluates each tree to decide whether it should be cut or left to grow for a later harvest and marks it with a ring of paint to guide the chainsaw operators. The forest is actively managed for an outcome that defers some immediate monetary gain but improves forest health and future profit. An Old Order sawmill owner in Ohio whose firm actively promotes selective harvesting mused about the difficulty of persuading landowners to maximize value at the next harvest rather than the current one: "We prefer to do a management cut, we're big believers in it, but it's still whatever the landlord wants to do. The problem is, when another buyer comes in and offers this amount of money, and he's taking *all* these trees [a diameter cut], and then [we] come in and offer *this* amount of money for *these* trees [management cut], that's just a big disparity, in the pricing. And a lot of people will see the big dollars for right now." Selective harvesting also requires foresters who mark the cut to know current timber markets, understand growth rates and habitat preferences of different tree species, and be able to assess tree health and recommend management practices that will encourage tree growth. Forest management requires a long-term vision of how the forest will look twenty years into the future.

Without direction from the landowner, Amish loggers are more likely to choose a diameter cut and to try to maximize economic gain from a timber job. One study in Wisconsin reported, "Many local residents [say] the Amish 'mine' the soil and ruin the forests. . . . They harvest timber without planning for forest regeneration or future management (of course many non-Amish do exactly the same)." The authors go on to note that Amish loggers practice "forestry like it's always been done."[9] Diameter cuts were the standard practice in an era when there was more forest than anyone needed. Amish loggers learned their skills from their fathers or grandfathers, who taught them the diameter cut. Though they know that a diameter cut makes the forest look bad in the short term, they may be unaware of the long-term impact on forest health, or else they downplay those effects. Since forests are meant to be harvested, they tend to see the natural cycles of growth and recovery as God's responsibility. One Swartzentruber sawmill owner told us that his Amish suppliers always had plenty of logs for him to buy. "As population grows, as there are more people, there are more trees," he said. "Somehow the Lord just keeps pushing 'em up. He has a formula for this, He's keeping track. We don't have to worry."

Most Amish buyers purchase timber from English landowners, who represent the majority of owners of private timberlands in Ohio. The Amish firms who practice management forestry report that 80 percent of their clients are English. On the other hand, Amish landowners rarely sell their timber to non-Amish buyers. When we asked why Amish landowners were not consulting forestry professionals or selling to non-Amish buyers, one Amish logger told us, "Well, the Amish are extremely—to put it kindly—thrifty. They have a big fear of being taken advantage of. You know, am I getting all I can for my trees?" He added, "And of course half of the Amish have a relative or a friend who's in the timber business, that's a factor." As a result, non-Amish foresters work mostly for non-Amish landowners. A consulting forester told us that he had almost never marked a harvest for an Amish landowner. Overall, though, only about 20 percent of timber cuts in Ohio are marked by a professional forester, who would guide the landowner toward a management cut. The other 80 percent of landowners harvesting their timber are less likely to learn about options for selective harvesting. And since about four-fifths of timber buyers in the central Ohio area are Amish, most harvests on these forests are probably geared toward timber production, not sustainable forest management.

Amish timber buyers and sawmill owners who understand and promote management-based logging are increasing, however, often as a result of contact with English forestry professionals. Some Amish sawmills and loggers see selective harvesting as an investment in future timber production and habitat enhancement, especially for deer.[10] While they would not describe it as being ecological, they would recommend it as good stewardship, and their efforts have begun to influence how other Amish loggers approach timber harvesting. An Amish forester who routinely practices forest management told us, "I've been buying timber for close to twenty years now, and in the last ten years we see a big change. More and more landowners are aware that they have the ability to get an every ten-, fifteen-, twenty-year income off of this ground." He said he was seeing more Amish logging firms talking about management cuts, but, he added, "I don't really know how they go about it. Are they really doing what they say they're doing? I don't know."

Another Old Order sawmill owner described how market pressures have altered his logging practices. He used to go into a site and cut everything with a diameter above sixteen inches. "I always felt terrible," he said, "'cause the woods looked horrible." But since the sawmill adopted a twenty-one-inch minimum cutting diameter, "now it's much nicer," he said. The twenty-one-inch minimum "leaves more for a later harvest, and we can reharvest on an eight- to ten-year cycle." He described this as an investment by his firm, since he wanted to encourage his customers to call him again, when they were ready for their next cut. This sawmill owner still practices a diameter cut, albeit less heavily than in the past. "You don't want to take them all," he said. "But if it's lost to a storm, that's money down the drain."

In the absence of a specific directive from the landowner, well-meaning buyers mark cuts in many different ways. There is always an element of uncertainty in an estimate. An Amish forester who is a staunch advocate of sustainable forest management told us, "One of the reasons why there's so much temptation in this,

I think, is because, when we're putting a price on a stand of timber, it's *always* a guess. 'Cause you can't look inside of that tree [to see if it is hollow or diseased]." Without advice from a forestry professional, the average landowner has no way to evaluate a bid. Stories abound about timber buyers, both Amish and non-Amish, who purposely underbid a sale and make a substantial profit when they market the sawn product from their mill. As one non-Amish forestry professional put it, "So it's very difficult to sort of gauge [whether the offer is reasonable], as a landowner. It's like throwing a dart at a board. I could talk to you for hours, on stories that I got from old timber buyers. . . . 'I paid that woman 17,000 bucks, and I harvested $90,000 worth of timber.' 'I paid that guy $100,000, I cut $500,000 of timber.' They are endless." And both Amish and English timber buyers are implicated in these lucrative deals. The lesson for the landowner is clear: get a professional estimate so you know what your trees are worth.

Amish timber buyers clearly incur risks in estimating a timber sale, but they also have abundant opportunities to make big profits. The price a buyer offers a landowner for his timber stand will depend on how many trees the buyer expects to cut, the way the loggers will remove the trees, and how honest he is with the owner about the trees' value. In general, most Amish timber operators see the forest principally as a resource to be utilized. The more trees harvested, the more value the sale has. It is a business transaction, with forest resources at the heart of the deal.

Cutting Trees

An Amish logging team of beautiful Percheron horses pulling logs out of a snowy forest is largely a thing of the past. A few Amish from conservative groups, however, still practice horse logging because of church constraints on technology use. Horse logging is one of the gentlest ways to take trees out of a forest, because horses' hooves compact the soil much less than a tractor. An Indiana sawmill owner described how his crew cut the timber to length on site and used a generator-run clam boom to stack the logs onto the horse-drawn wagon. But horse logging cannot meet the economic demands of commercial harvesting. Instead, most Amish timber crews use modern tools, from chain saws to bulldozers, to cut forests. Gasoline- or diesel-powered skidders make it relatively easy to extract the timber, but they also have destructive consequences for the forest habitat. The basic tradeoff for an Amish logging crew is between time and money. Reducing machine damage means working more slowly and carefully. However, the economical way is to harvest quickly, get the logs out, and move on to the next job, leaving the forest to recover on its own.

Despite prohibitions against driving private vehicles, most Amish workers are

Skidders like this Caterpillar with treads, rather than tires, do less damage to forest soils as they push the cut trees down and prepare to drag them from the woods. Photo by Marilyn Loveless

allowed to operate tractors, dozers, and log-loading grapples in the forest. The crew begins by cutting access roads into the site. These roads will carry heavy tractor traffic, compacting the soil. The ruts left behind act as channels in heavy rain, eroding to deep gullies after logging is done. Compacted soil also slows vegetation recovery, allowing further erosion. Erosion from logging roads and damage to streams negatively affect water quality, as siltation can kill aquatic invertebrates and kill the fish populations that feed on them. To reduce damage, roads within the forest should be as short as possible. Loggers then use chainsaws to fell the marked trees. Skilled fellers can drop a fifty-foot oak into a narrow space where it will do as little damage as possible to surrounding trees. The downed tree is then cut into manageable lengths, and limbs and tops are trimmed off.

Downed, trimmed logs are pulled out by an Amish-operated log skidder, a large-tired tractor with a grapple arm on the front and cables on the back that attach to the felled logs. As the skidder makes its way up slopes, through downed tops and limbs, the tire treads tear up the forest floor, leaving deep ruts and tracks. A few

Amish crews have track skidders, with continuous tread rather than tires, which are much less destructive. The logs are dragged to the landing, often a distance of several hundred feet. The landing, or yard, is a cleared area where logs are stacked to be loaded on trucks for transport to the sawmill. Dragging logs out to the roads also creates deep ruts in the soil. Unless the crew works carefully and deliberately, logs can scrape or gash the lower trunks of trees that were not marked for cutting, killing them or reducing their growth and future value.

Stream crossings are especially vulnerable to logging damage. Logging along stream edges lets in more light, increases water temperatures, and promotes algal growth, which alters the stream ecology. Even though the vegetation grows back after a harvest, there are long-term ecosystem impacts that are difficult to assess. As a result of governmental regulations, Amish logging crews have become increasingly attentive to the damage that can come from dragging logs across forest streams. An Old Order sawmill owner and logger explained that "for crossing streams, we use steel bridges. We don't just drag logs through major streams where erosion can be a problem. And there's not very many people doing that any more. You don't have to go back too many years to where, that was a normal thing. Even Amish loggers are [using bridges], more and more." When the last tree has been taken out, a careful logging crew will remediate the site. Roadbeds must be graded with a bulldozer and reseeded to reduce erosion. The crew should install water bars, angled dikes that direct flowing water off the graded surfaces into less-compacted soil along the sides of the roads. The removed treetops and other waste wood should be chopped up to make the forest more easily passable on foot. Doing all this takes time and requires training and care.

This suite of forestry practices—careful felling and skidding, avoiding damage to trees, grading roads, reseeding, constructing water bars, using portable bridges across streams, and imposing a setback along stream edges—are collectively known as best management practices, or BMPs. Ideally, they should be followed in every timber harvest to prevent long-term ecological damage to forest habitats. To implement them effectively, Amish logging operations need training. Most states have workshops where loggers can learn these practices.[11] BMPs lead to large-scale, long-term benefits for the health of a forest, but regulations regarding use of BMPs in timber harvesting vary widely among states.[12] In West Virginia, BMPs are required, and inspections follow timber jobs to ensure compliance. As a result, said one Ohio Amish mill owner, "a lot of operations just left West Virginia." Indiana also has strong regulations that protect water quality during logging, as one Amish sawmill operator near Goshen attested. "If you damage a stream, in a way that disrupts the fish or the stream," he said, "there's enough music to face that we want no part of that." Pennsylvania requires permits for various aspects of the harvest, specifically to enforce BMPs, but some states, like Ohio, have no regulatory oversight of logging on private properties.[13] Both Amish and English loggers in Ohio are only required to do those things that are specified in the contract.

In general, Amish logging operations have not been at the forefront of careful logging practices. Some Amish logging operations implement BMPs voluntarily, either because they represent good stewardship or because they result in a more satisfied landowner.[14] But if BMPs are not required, a logging operation will not implement them unless the landowner makes them a contractual requirement. As one Amish logger told us, "We have this challenge. The landowner wants every penny out of his trees—which we totally understand. But we want to do as nice of a job in the woods as we possibly can. We want to chop the tops down, make sure the roads are graded in nice, and seeded in, and everything's done right. But there's a limit of what we can do to still stay profitable. We can't pay more than everybody else, and do the job that we'd like to do." A non-Amish forestry professional in Ohio agreed that unless loggers have gone through a course to learn BMPs, "quite frankly, most don't care, because BMPs take time and effort to do." Even when loggers do go the extra mile, their efforts are not always noticed or appreciated. One Ohio Amish sawmill owner with an overwhelmingly English clientele told us, "We try to maintain the forest. We explain what we are doing. But only about twenty percent of landowners really care." Nevertheless, in the modern forestry industry, failure to use BMPs would be viewed as environmentally irresponsible, resulting in harm to the landowner, who is left with a degraded forest landscape.

Sawing Lumber

Another pivotal but less visible role of Amish in the forest products industry is as owners or operators of sawmills. In areas with abundant forest potential, timber milling is big business. Sawmills range in size from small sheds in the backyard, with a single circular blade, to large, computer-controlled operations with multiple saws and mechanical systems for debarking and moving the logs and lumber through the mill. Many Amish sawmills do some or all of their own logging, hiring procurement foresters or timber buyers to find tracts of trees for harvest. They may also buy timber from brokers or other mills. Some mills specialize in certain species or in certain kinds of wood products. In all these operations, Amish businesses must use non-Amish drivers to move men, machinery, logs, and timber to their destinations. Because of their strict limits on technology, conservative Amish usually

In some situations, Amish use portable sawmills to cut lumber in the woods, rather than hauling it to a distant mill for processing. Photo by Doyle Yoder

operate small, one-blade mills that run on a gasoline motor. Frequently, their older male children help with running the mill and stacking and carrying timber, and neighboring Amish families use wood scraps for firewood. Swartzentruber Amish may saw logs for construction lumber or build sheds or deer blinds. But they mostly focus on generic items that do not need smooth boards and can be sold as smaller units, such as pallets, posts, railroad ties, and garden stakes.

The red oak for our hypothetical dining-room table, however, would be milled by a larger operation, capable of various types of saw cuts and of finishing the rough boards by planing and smoothing. Many Amish sawmill owners belong to the second or third generation, with a family tradition of wood processing. When logs arrive at the mill, they are usually sorted by species on the log yard. Veneer logs, which bring exceptionally high prices, are separated and sold to specialized veneer mills. Sawmills typically focus on grade lumber (length boards), but some may produce specialized "dimensional" pieces that have a market in various secondary woodshops. Another, more recent wood-products market was described by an Amish sawmill owner: "The big thing right now, actually, is barrel stave logs—white oak logs—for whiskey and wine barrels. What I heard is, they can't get enough logs. One big thing is, the Chinese have developed some form of a middle class, and they're big drinkers. But generally it's the smaller logs, down to eleven inches, because the bigger, good logs get sold for veneer." As this owner made clear, some Amish wood producers are increasingly integrated into national and international demand cycles. They are good at identifying niche markets and are attentive to supply and demand, even if it takes them far afield of the local furniture market.

After emerging from an Amish sawmill, the lumber must be dried before it can be used. Some sawmills have dry kilns, but others sell the green timber to an intermediary, who does the drying and takes care of the ultimate distribution. Wood is stacked in large heated and ventilated rooms, where it is allowed to dry for two to four weeks, an energy- and labor-intensive process. Restrictions in some Amish communities on the use of fuels like natural gas make it difficult to run a dry kiln. While kiln-dried lumber brings a slightly higher price to the wholesaler, it adds complexity to the mill operation.

Sawmills must follow OSHA (Occupational Safety and Health Administration) and EPA (Environmental Protection Agency) regulatory rules. Amish business owners often grumble about "excessive" safety regulations, but they generally try to comply.[15] They do not want their workers injured, nor do they want to risk fines and negative publicity from potential pollution. Workers wear dust respirators and eye protection, but the Department of Labor waived the hard-hat requirement for Amish employees in 1972.[16] The EPA regulates disposal of bark and wood chips and sawdust, most of which can be recycled into mulch or horse bedding. But small Amish sawmills sometimes don't have the volume of sawdust or a nearby market that would make recycling practical. In one Kentucky settlement, an outdoor sawdust pile from a small Amish sawmill was leaching into a creek, turning it black, and impacting water quality in a nearby lake that was popular for recreation. Local officials "threatened to shut it down," the bishop of the church district told us, but ultimately a compromise was reached to more effectively contain the sawdust pile. In addition to the problem of sawdust disposal, if the wood at a sawmill is treated with a chemical protectant, the mill must safely dispose of the drippings and other materials.

An Old Order sawmill owner in the Midwest showed us around his large sawmill and log-home business, which exemplifies the complexity of the wood-products industry. A saw shed on site processes the white pine harvested within a one-hundred-mile radius of the sawmill and destined for log-home construction. This Amish mill owner also buys hardwood timber, contracting the logging to a mostly non-Amish firm. The hardwoods are processed in another, nearby sawmill. The non-Amish forester who oversees the logging recommends forest management to the landowners. "He talks to the homeowner and tells them what they should do," the owner said. "But it depends on the homeowner. Sometimes they have a need, and they want to cut."

Trimmed, squared-off pine logs are dried in one of three lumber kilns, with a total capacity of 94,000 board feet. The kilns are run by diesel- or propane-gas motors that power fans that circulate hot air through the rooms. The business recycles sawdust from the mill for use as horse bedding. The trimmed edge slabs from the pine logs are used for shingles and small boards, to avoid waste. The precut log-home materials are then shipped nationally and internationally and assembled on site by local contractors. This mill owner called his timber "carefully harvested," and he did not want his loggers to "mine" the forest. He was critical of those who harvested "down to the toothpicks" and then sold the denuded land as lots for homebuilders. In contrast to mill owners who make little effort to minimize waste or to monitor their timber sources, this mill owner emphasized selective harvesting, as well as reducing and recycling wood waste. Like non-Amish mills, Amish sawmills are a diverse set of enterprises, privileging economics over the environment in some instances but embodying environmental considerations in others.

Using Wood

The most visible step in the Amish wood-products supply chain is the furniture store that sells that red oak dining-room table. Many tourists love Amish-made furniture. Its solid-wood construction gives pieces durability and beauty. For English customers, artisan-made hardwood furniture epitomizes the perceived connection of the Amish with nature. Wood is warm, traditional, renewable—qualities that evoke its closeness to the natural world. An Amish employee of an upscale Amish furniture retailer said, "Customers like that [the furniture] is built locally. They associate it with hard work. They envision that it is hand made." But, he added, "most of them know that we have tools to do it." Furniture is the best-known Amish wood product, but Amish woodworkers make hundreds of other items, from clocks to toys and from dining-room tables to caskets. Many Amish wood shops take pains to make smaller items like magazine holders, garden art, and Lazy Susans, which they call "trunkables," since they are small enough for a visiting tourist to put into a trunk.

Wood is also used in construction. Well beyond the iconic "barn raising" of lore, Amish contractors and crews design and build homes, garages, barns, and other structures for both Amish and non-Amish customers. Amish construction firms

In some settlements, nearly half of the men work in the wood-products industry, including many in furniture shops like this one. Photo by Doyle Yoder

employ multiple teams of workers. Van drivers pick them up at their rural homes in the early hours of the morning and drive them ten, fifty, or even a hundred miles or more to do roofing, siding, cabinetry, flooring, or other construction-related jobs. Some work teams even travel out of state for a week or more at a time. Though the structural components of a building are not made from hardwoods, the interior details, such as cabinets, doors, stairways, bannisters, and flooring, often are.

While furniture manufacturing is an economic engine of the Amish community in its own right, Amish businesses are integrated at every level of the supply chain.[17] In the Holmes County settlement, more than 85 percent of the lumber used by Amish furniture firms comes from local distributors and sawmills.[18] This ready source of local timber and the many businesses who carve out specialty roles supplying the furniture industry make the whole wood-products sector more efficient and give it a competitive edge that most small manufacturing businesses cannot match. That red oak table that goes back to Chicago in a visitor's van most likely came from local timber, local sawmills, and local woodworkers. The same process occurs in all the large Amish settlements in the Midwest, most of which are within 150 miles of several large population centers. Though not all Amish settlements are equally close to large, wooded areas, the wood-products supply chain is alive and well in any Amish community that has access to a forest resource.[19]

Furniture manufacturers, like sawmills, must comply with health and environmental regulations. For hardwood manufactures, the principal concerns for both OSHA and the EPA are sawdust disposal and management of overspray from the stains and finishes applied to products. Finishing requires ventilation and a way to collect overspray, which can often be reused to reduce waste. Most Amish furniture-shop owners see the necessity for these measures, though compliance varies. We visited one furniture shop that did about $5 million in annual sales and had installed a state-of-the-art ventilation system. Yet at a much smaller shop, we witnessed several Amish teenagers applying sprays with minimal protective gear, and we watched one Amish cabinet refinisher spray varnish in a client's kitchen without using any protective gear at all. These examples suggest that at least some Amish woodworkers still minimize the health and environmental risks associated with toxic chemicals.

The timber industry in Amish settlements in Montana, Wyoming, and Colorado looks quite different from that in the East because the natural and social landscape impose different constraints. Western forests are made up largely of conifers and lack the hardwoods that nourish the eastern woodworking industry. In addition, land-ownership patterns differ from those in the East. Some western forests are part of large landholdings owned by giant timber companies like Weyerhauser and Plum Creek. But the standing timber in lands owned by the National Forest Service and the Bureau of Land Management dwarfs the timber that might be available from private holdings. Timber sales on federal lands are planned by government agencies, but buying, logging, and milling are done largely through contracts with big companies. Another dynamic is that logging in national forests is under continual scrutiny by environmental advocacy groups, and timber sales may provoke controversy. Amish in the western United States must navigate these different ecological and political contexts if they want to tap forests for timber.

Western Amish woodworkers seek out niche industries and manufacture materials that, because of their unique character, can be marketed effectively. In St Ignatius, Montana, one business owner buys small-diameter lodgepole pine from logging firms who bid on timber sales in the surrounding region. The mill produces rails, posts, and hand-drawn (debarked) lodgepole pine poles that are used for rustic furniture. Another mill uses cedar from the area to make decorative shingles. In Rexford, Montana, a woodworking shop buys lodgepole pine poles and stumps from loggers who have permits to salvage beetle-killed trees on federal land. They saw poles, but they import all their other sawn timber and components from mills in Oregon and other western states to make their popular rustic furniture. An Old Order Amish sawmill in Westcliff, Colorado, cuts local fir, spruce, and pine for lumber, mostly for local markets. The mill also buys or harvests dead aspen poles, which it ships to Indiana, Ohio, and Pennsylvania for use in manufacturing rustic furniture. This sawmill owner also operates a retail furniture store, selling hardwood furniture bought, custom built, from furniture manufacturers in the Holmes County, Ohio, settlement. This latter case demonstrates that Amish woodworking businesses are not only vertically integrated but maintain horizontal linkages across considerable geographic distances.

The Renewable Forest

When the Amish first arrived in "Penn's Woods" (Pennsylvania), they found a landscape dominated by dense, almost continuous stands of white oak.[20] Trees were too abundant to be considered a resource; more probably, they were seen as an impediment to human livelihoods. In eighteenth-century America, human populations were small and forests were vast. People cut trees in the easiest, most efficient way, clearing land for agriculture and making use of lumber without concern that the resource was limited. Only after the Civil War did the human potential for forest exploitation finally result in an environmental catastrophe. Increasing demands for lumber gave rise to industrial-scale logging that decimated the pristine forests of the Northeast and the Midwest over the next fifty years.[21] Driven by new technologies, commercial loggers clear-cut huge tracts of land, with no thought to forest regrowth or to the erosion, flooding, and fires that followed them.[22] As the century closed, only 4 percent of the old-growth presettlement forests remained.[23]

Out of this calamity the idea of forest stewardship was born.[24] The model has evolved into what is now known as "sustainable forestry," the selective harvesting and environmentally sound management of forest resources. In the meantime, the eastern forests have regrown. But forest tracts are smaller, more patchy, and different in species makeup than they were when the first Amish families settled in Penn's Woods. Many forests are young, having returned to woodland from farms abandoned over the last decades. The human population continues to grow, and there is increasing demand for all sorts of forest goods. At the same time, there is a growing recognition that the planet needs intact, healthy forests, which not only grow the trees of the future but provide habitat for myriad species, protect soil from erosion, filter water, store carbon dioxide, moderate stream flows to reduce flooding, and help to purify polluted air. For these things to occur, forests need to remain forests, and environmentally sound forest stewardship means using forest resources without eliminating the forests or impairing their function.

Amish who work in the wood industry in the twenty-first century thus operate in a very different cultural, political, and ecological context than their parents or grandparents did. As Lynne Heasley and Raymond Guries note, "American forests have come to represent more than the biological sum of their trees. They are the material and symbols society wields in its debates over nature, the environment, natural resources, and property."[25] Some Amish sawmills and furniture businesses have changed with the times and implement best management practices across the board. But many others have been slow to change. They value the forest for its timber and remain unsentimental about cutting trees. An Amish sawmill owner told us, "I'd say it's a crop as much as any other. Trees. We might look at it that way, more than some." This comment helps to explain why heavier cutting practices are not uncommon among Amish crews. Another Amish sawmill owner declared that most Amish are more aware of "what's happening in their woods" and "what makes the wood grow" than non-Amish landowners, who appreciate the woods more for aesthetics. In his view, the Amish understand forest processes, because they still look at the forest largely in terms of human use and are more focused on the income a forest can provide.

Many Amish also approach the forest with the understanding that this world is transient and they are only passing through on their way to a better place. When we asked how Amish loggers in Indiana felt about regulations governing logging, an Amish sawmill owner replied: "Well, the Amish have a tendency to not have a horizon that's that far out, because of how they separate themselves from the world." An Ohio Amish forester offered a similar explanation for why the Amish might not actively try to protect natural resources. "I've always thought that one of the reasons is, because the Amish look at this as a temporary thing," he said. "A person's belief of what reality is, and of what this earth is here for, will greatly dictate how we use it. Even though we want to take care of it, it's not going to be quite as much of a priority, if we know it's temporary." A preservationist perspective, as one Amish man told us, is not something the Amish embrace. But many non-Amish logging companies are no different. They focus on timber production, cutting heavily and failing to remediate the disturbance that results. As one Amish sawmill owner said, "Like any other business, there's good and bad. There just is."

Over the past forty years, a movement to "certify" properly harvested timber, especially from tropical forests, has taken root. Certification sets standards for good forestry practice, and customers who purchase certified products are guaranteed that the wood was grown, harvested, and processed in compliance with strict ecological and social criteria.[26] Certification requires changes in practice and demands extensive documentation and recordkeeping, but it represents a strong environmental commitment. The success of this movement, however, depends on market demand for these well-managed forest products. And unlike organic produce, green furniture has proven to be a hard sell. Only a few Amish firms have enrolled in certification programs sponsored by professional business associations, in hopes of attracting more environmentally concerned customers. But one high-end Amish furniture maker said that maintaining certification has been a huge drain on his time. Though he tells his customers about the sustainable initiatives his business has implemented, "they don't care. . . . We've even gone so far as develop a water-based stain system that works really nice, but nobody's willing to pay the fifteen percent up-charge of what it actually costs to implement." Without consumer demand for green forestry products, Amish businesses are unlikely to pursue certification programs.

The idea that a product is Amish-made may actually carry more cachet with consumers than a green-certified label. To many outsiders, "Amish" connotes quality, craftsmanship, care, and hard work. Many Amish refrain from marketing their products as "Amish-made." But the lines are blurring. An Amish man who works for a store selling Amish furniture said, "Our slogan is 'Built by the Amish.' It's a big marketing tool." He told us that his boss, who is not Amish, had been advised to hire some Amish men to provide visual reinforcement of the Amish origins of the products sold. A similar Amish advantage might apply in timber buying and logging. The idea that the Amish are trustworthy may work in their favor. A non-Amish forester said, "I think that people just have a view of Amish, that they're just better, because they're Amish. It probably has something to do with religion. They're highly religious people, they must be good." Tourist marketing for Amish communities is sometimes even more unguarded in its use of the Amish label. As noted at the beginning of this chapter, the tourist-oriented article "The Green Machine" touts the eco-friendliness of the Amish wood-products industry at every step. Our own observations of the supply chain and our interviews with Amish and non-Amish in the wood-products industry provide a much less sanguine appraisal. Yet if non-Amish customers are already inclined to see Amish wood products as ecologically motivated, tourist-aimed articles like "The Green Machine" provide them with apparent evidence.

Tinkering with Creation

Alternative Animal Breeding

If you ask him about his herd, a farmer whom we will call Howard can recite his animals' lineages, of both sires and dams, going back three generations. The breeding records are filed in binders in his "office," a small alcove off the kitchen. All his animals have been DNA genotyped to verify their lineages for potential buyers.[1] He carefully matches males and females to produce high-quality offspring, and the newborns attract a lot of attention. When his children were young, they loved to bottle-feed them. He buys specially formulated diet supplements, and ear-tags each animal for ready identification. His farming operation has been a good change from his previous woodworking job. Now, he can be at home, and his family has actively participated in the business. Howard has been savvy, bringing in new animals from strong genetic lines, and as a result, his animals bring high prices. So he's making more than enough to provide his family with a comfortable living.

But what is Howard raising? They aren't cattle—or any animal you might expect to find on an Amish farm. Howard is a whitetail deer farmer. He belongs to NADeFA, the North American Deer Farmers' Association, and all his animals' genotypes are archived in the North American Deer Registry. He's one of a growing number of Amish who earn their living by farming in ways that push the boundaries of traditional farm culture.

Animal-breeding operations are a new face of farming among Amish adapting to the realities of the twenty-first century. Family friendly and requiring little acreage, they appeal to Amish across the country who seek a business that resonates with their rural, agricultural identity. But if animal husbandry is an Amish stand-by, its twenty-first-century expressions are explicitly designed for economic profit. Sometimes these endeavors scale up a hobby like breeding dogs or exotic animals in order to meet a wider market. On the other hand, Amish horse and deer breeding focus on specialty markets, involving careful attention to pedigrees and strong selection for animals with sought-after traits. The goal is not just to produce offspring but to alter nature, to shape the characteristics of offspring as a way of adding value—sometimes considerable value—to the progeny of a cross. High-end horse breeding has challenged the traditional role of horses in Amish life. Whitetail deer breeders have taken a wild species and transformed it into an exaggerated and high-priced commodity for private hunting ranches. All these endeavors represent extensions of traditional husbandry that are culturally appealing and tap into potentially lucrative markets.

Animal ownership is a pervasive element of the Amish rural lifestyle. Businesses based on animal breeding, however, may take owners in directions that push the limits of church doctrine. Breeding is largely driven by demands from outside the Amish community and has led the Amish to adopt new technologies, including DNA markers, artificial insemination, and internet marketing. These activities demonstrate how the Amish use animals as commodities within the larger economic system and provoke conflicts about whether these livelihoods are consistent with Amish values. In this chapter, we examine these nontraditional animal businesses, asking how Amish practice these enterprises and what they suggest about how the Amish think about nonhuman species, both domesticated and wild, as elements of nature.

Exotic Animals

The Amish share a human fascination with species that are rare, impressive, or beguiling. Often this starts as a childhood interest in animals. Day trips to nearby zoos are common outings for Amish families. An Amish bird breeder enthused, "We would go to see peacocks [at the zoo], and I was like, 'Oh, I gotta have one of those.' " Collecting and breeding exotic species may start as a hobby but can quickly become a business, allowing an Amish breadwinner to balance a "lunchpail" day job with something like farming in the off hours.

For some, raising exotic species is purely an economic venture, but for others it is a labor of love. Owning rare animals admits you to a circle of people who share your fascination. Rare pets of any sort also provide an outlet for interest and affection.[2] A young Amish man who specialized in breeding Asian pheasants and doves told us, "It takes special people to raise special birds." This New Order man and his brother housed nineteen species of exotic pheasants in long rows of hand-built, carefully landscaped cages and had converted a cast-off mobile home into a breeding facility, raising chicks of more than a dozen bird species.

Exotic animals, from parrots to pot-bellied pigs, can bring hundreds or thousands of dollars each. "The Amish have their finger on the pulse of what's going on in the community," a veterinarian in an Amish settlement told us. "For a while it was ostriches, then emus, and llamas, and they've gotten out of all of that because you

This Amish-owned aviary in a settlement in the Blue Ridge Mountains of Virginia housed a number of exotic bird species and had business relationships with several prominent American zoos. Photo by David McConnell

can't get money from it anymore. But whatever people are willing to buy they're willing to raise." From a conservation perspective, however, owning and selling exotic animals presents an ecological contradiction. As collectors' items, they become caricatures of their wild and free origins. Breeding exotic animals not only monetizes them; the demand helps drive a market for illegal capture of animals in the wild. The Convention on International Trade in Endangered Species (CITES) restricts importation of many species. Nevertheless, illegal animal smuggling is widespread and jeopardizes native populations of sometimes threatened animals.[3]

The regulatory environment for ownership of exotic animals is fragmented, and most Amish exotic ownership requires no permits. While owning large, dangerous animals such as big cats and bears is prohibited in many states, smaller animals are completely unregulated. Amish vendors and breeders must follow USDA regulations in some circumstances, and most are compliant, but lack of inspection and oversight often permits abuses. People who support ownership of exotic animals argue that animals are property and that banning them would violate a citizen's constitutional rights. Those who oppose the exotic market argue that undomesticated animals with special habitat, health, and behavioral needs cannot be kept humanely as pets in captive situations.

The best window into the exotic-animal business is through "alternative animal auctions" in rural communities, where the buyers and sellers include both Amish and non-Amish traders. The Ohio Alternative Animal Auction, one of the largest in the country, is held three times a year in Mt Hope, Ohio.[4] Similar auctions take place in Topeka, Indiana, and in the Lebanon Valley of Pennsylvania, as well as in non-Amish areas. The Mt. Hope sale list includes zebras, antelopes, camels, foxes, capybaras, hedgehogs, and birds, from canaries to macaws. One Amish man we spoke with was selling a silver fox. Another Amish family had brought young pot-bellied pigs to the sale in plastic under-bed boxes closed with bungee cords and were busily trying to keep the piglets from escaping. These auctions attract the scrutiny of animal-rights activists, who compile undercover reports, available on the internet.[5] The auction, for its part, advertises its compliance with USDA regulations regarding health certificates and animal welfare.[6] Most Amish would not condone blatant mistreatment of exotics under their care, but the animals at the auction are valuable commodities. Some may eventually become pets, and others may end up in exotic-animal breeding operations, but all of them will live out their lives far from their native landscapes, sold to a public that finds them appealing.

Dog Breeding

Small-scale dog breeding has long been common in rural communities, but in the last twenty-five years it has been transformed into a livelihood for many Amish families. By 2006, dog kennels specifically devoted to breeding puppies were more numerous than dairy farms in Holmes County, Ohio.[7] Dog breeding is also a major business in other large Amish settlements. For some Amish, dog breeding supplements their primary income, but many have made it a full-time occupation. Some breeders sell directly to customers, but many sell their puppies to a broker, often Amish himself, who markets the dogs to pet stores or to internet outlets. The dog business is driven by demand from non-Amish buyers for purebred or designer hybrid puppies. Fashionable puppies can bring hundreds or even thousands of dollars each, so a large dog kennel provides a comfortable income for an Amish family. A successful horse breeder in Ohio declared, "These kennels, it's unreal how much money they make."

The Amish tendency to see dogs as livestock, as part of the natural world created for human use, is increasingly at odds with the way many non-Amish have come to view animals. A veterinarian in an Amish community told us, "There's been a huge shift in animal-human relationships in the last twenty or thirty years, because most people have moved away from the farms. When most people think about animal-human relations, it's a pet situation, so lots of emotions involved. Amish with their animals aren't nearly as sentimental as we are. It's not to say that they're not going to care about their pets, but you know, it's a business." As a business, dog breeding has quickly spread among Amish families looking for alternatives to farming. In 2010, as the Ohio legislature considered legislation to regulate commercial kennel operations, the Amishman Ervin Raber, of Holmes County, a longtime, vocal advocate for dog breeding, estimated to one reporter that "60 to 70 percent of the families in his area stake their survival on income from breeding."[8] While such numbers vary by region, raising puppies has clearly become a major activity in many Amish communities.

When commercial dog breeding began among the Amish, there were virtually no regulations. The Amish housed dogs in their barns or in homemade kennels, bred them, and sold the puppies, no questions asked. Kennels were not inspected unless a neighbor filed a complaint with the local animal warden. One Amish breeder recalled, "The regulations came from the puppy mill talk." Over the past twenty years, dog breeding has increasingly drawn criticism, as advocacy groups publicized examples of poor health or problematic care for dogs in kennels and for puppies sold to pet stores. While most dogs were not actively mistreated, they didn't get the personal attention that non-Amish owners give their pets. The Humane Society of the United States (HSUS) and other animal-rights organizations, as well as a 2008 "Oprah" documentary, focused national attention on "puppy mills," which prioritize profit at the expense of animal welfare. With respect to the Amish, the animal activists' outrage was directed not only at the deplorable conditions in some kennels but also at the irony that a people widely seen as gentle and religious were treating dogs in ways that seemed contrary to Christian values.[9]

Animal care among Amish dog breeders, however, varies along a spectrum. At one end, as Kraybill, Nolt, and Weaver-Zercher note, "dominion means squeezing benefits from animals with the least amount of cost. . . . These Amish farmers regard nature as little more than a means to an economic end."[10] At the other end are Amish dog breeders who provide adequate food and water, veterinary care when appropriate, solid cage floors, good ventilation and heating, clean quarters, access to outdoor runs, and regular human interaction. The Amish breeding facilities we visited were simple and utilitarian but generally clean, and several breeders stressed the importance of socializing their dogs. One Old Order man told us that he uses

Amish dog breeders are attentive to the demands of the market and specialize in popular breeds or cross two breeds, such as golden retrievers and poodles, to create a hybrid known as a goldendoodle. Photo by Marilyn Loveless

corn syrup to help puppies get over their fear of him. "If you give 'em syrup, they get real social," he said.

Amish dog breeders tend to specialize in particular breeds or hybrids that are sought after by English customers. An Old Order dog breeder whose brother bred bulldogs looked at the competition in the area and decided he could fill a niche for Havanese. Another specialized in a designer hybrid known as a Pomsky, a cross between a Pomaranian and a Siberian Husky. Breeding a big dog like a Husky with a small Pomaranian, however, is no easy task, and kennel owners have largely trained themselves in the science required to conduct "assisted breeding." One Old Order dog breeder has family members help him as he uses a mail-order kit to test for ovulation and then collects semen by putting the male dog on his lap and stroking it. He even bought a semen extender and a microscope so that he could determine whether the semen was still alive. "It's pretty cool to see it swimming," he commented. He inserts the semen into the female and holds her rear end up for five to ten minutes. "The first time I did it, it was a mess, and I had to call my wife to help, but now I can do it by myself," he said. Then he waits for a week or two to see if the pregnancy takes. This trial-and-error process allows Amish dog breeders to become proficient in the techniques they need to meet the demands of their customers for designer dogs.

Over the past decade, several midwestern states have passed legislation intended to tighten licensing and implement better standards of care. One Pennsylvania Amish man we spoke with argued that the Humane Society had targeted breeders, making breeding very difficult. "I can see taking good care of a dog," he said, "but what they were having to do, I don't know. If you have under twenty-five dogs, you can avoid the regulations, so that's what happens. Or they moved their operations to Ohio." Ohio's 2013 Commercial Dog Breeder Law, written with the active participation of the Ohio Professional Dog Breeders Association, was declared a victory against abuse. But activists claim that it does little more than require breeders to follow the minimal standards of care already defined in the USDA Animal Welfare Act.[11] And some dog kennels still are not licensed either because they fall below the statutory size for high-volume breeders or because they stay under the radar in various other ways.[12] While licensed breeders are subject to periodic inspections and must follow state regulations on kennel conditions and animal health, small-scale or unlicensed breeders are unregulated. In spite of high-profile laws passed in some states, reports of mistreatment and abuse still emerge in the press.

Some Amish dog breeders, however, see value in the stronger regulations. One man observed: "For our furniture business, being Amish is good. But for puppies, [English] kind of like get a red light, because they think puppy mills are all Amish.[13] The Amish name with respect to puppies is more negative than positive, but it's better than it was." He continued, "The new rules and regulations made the dog business better. It winnowed out the bad people, and left the rest stronger. It prompted our local group to come together, step it up a bit, and do it right." Most state-licensed dog breeders have, in fact, moved toward standards of care that would meet non-Amish expectations. Some Amish that sell directly to their customers have exceeded the minimum standards, building state-of-the-art kennels with indoor and outdoor dog runs, solar and gas-driven heating and fan systems, and a strong emphasis on health and disease control.

The Amish increasingly realize that bad publicity hurts their bottom line, and they hope that new regulatory laws have improved public perceptions about Amish dog farms. But the differences in values are hard to overcome. In 2014 a young Amish man in Westcliffe, Colorado, applied to his local zoning board for permission to open a dog kennel. Animal advocates from nearby Denver attended the hearing, raising the specter of a puppy mill in the neighborhood. The pressure eventually pushed the man to withdraw his application.[14] Another Amish man in Colorado told us that breeders in Amish communities "try to stay under the radar" by keeping their operations small so they don't require zoning variances or other permits that go into the public record and "cause a ruckus."

Amish kennel owners are usually unapologetic about their business, but their Amish neighbors are sometimes privately uneasy about dog breeding as an Amish lifestyle. One New Order man said, "The animal rightists do have legitimate claims in some cases that some facilities aren't up to par, their practices aren't humane, and our people recognize that more and more, that it's not excusable just because you're Amish. We sort of look at people doing this for big financial gain as doing it for the wrong motivation. It's becoming more of a questionable activity than it used to be at one time." An Old Order man echoed this sentiment. "Dog breeding—that wouldn't be for me. I do not advocate puppy mills. Is there an opportunity there? Absolutely. But I wouldn't feel right. If we go back to our roots, this isn't what we want." Although dog breeding is widespread in Amish communities, it is not uniformly embraced by all Amish households.

As a result of the negative pressure they experience, dog breeders in Amish settlements keep a low profile. Their kennels are often windowless and set back from the roadways. We had a difficult time finding dog breeders willing to talk about their business. One man we contacted said, "I'd love to talk with you, but I've been advised not to talk with anyone." Dogs have become an emotional lightening rod for conflicting world-views about the nature and use of animals. The fuzzy puppies that Amish kennels produce in bulk are ultimately adopted by non-Amish customers who treat those dogs like pampered children. The conflict between the Amish dog breeders and the animal-rights activists exemplifies the friction that occurs when a relatively self-sufficient, strong-minded community is forced to negotiate boundaries with a surrounding culture whose values they have long been able to selectively ignore. To the degree that Amish businesses depend on non-Amish markets and profit from the values reflected in those markets, they will continue to face scrutiny from those same markets and be called upon to engage in difficult conversations across starkly opposing moral terrains.

Horse Breeding

The Mid-Ohio Memorial Catalogued Trotting Sale, started by a group of Amish horse breeders, is an annual event at the auction in Mt. Hope, in central Ohio. In 2016, its fourth year, there was a waiting list for inclusion among the 130 registered

Standardbred horses listed in the red, white, and blue catalog. Entries give the horse's description, pedigree, and race winnings, along with short, catchy descriptions to entice buyers. "Nice driving mare with a good pedigree." "Very well-mannered and a sharp driver." "Mile eating machine that is always fancy."[15] As the phrases suggest, these trotters are animated horses, and a spirited, stylish horse in front of your buggy sets you apart in the Amish community. But horses sold during the event are destined for the buggy harness, not the racetrack. And individuals, not just breeders, can participate. The sale is also one of the biggest benefit events in the local Amish community, contributing more than two hundred thousand dollars toward medical expenses for Amish families in need.

The overwhelmingly Amish crowd watches from bleachers alongside an oval field as each horse circles the track pulling an open, two-wheeled harness-racing cart called a "sulky." The auctioneer's patter is broadcast over the arena. In atmosphere the event is closer to a fair than to a horse race, but the money that changes hands is not popcorn change. Given the central role of the horse and buggy in Amish culture, horse breeding might appear to be the most natural of Amish pursuits.[16] But this seemingly innocuous event has led to a lively, sometimes contentious debate within the Amish community about how the horses are bred and the money they command. The strictly utilitarian Amish approach toward horses is being challenged by values that have diffused into Amish culture from the outside world—the world of the show ring and the racetrack.

Historically, horses were for pulling things. The Amish bred buggy horses to be well behaved and "Traffic Safe and Sound." Draft horses were selected for strength and teamwork. Amish values dictated that a horse should be plain, hardworking, and functional. But over the past twenty years, Amish horse breeding has taken a new direction. Horses have become a hobby, a source of recreation, and an opportunity for investment by some wealthy Amish. One Amish woman, watching her children ride in at the end of a benefit trail ride, told us that when she was young, children had ridden buggy horses. But riders now want Saddlebreds and Morgans, which have a smoother gait. "It's much nicer to ride a pedigreed horse that is trained for riding," an Amish trainer told us. Said another Amish man, "It's just a lot more fun to drive a horse that has 'action' and wants to go."

This new interest in aesthetic qualities has transformed horse breeding for some Amish farms. "Horses were always a passion for me," one Ohio Amish breeder told us. He said that when he had started out, "trying to live by breeding and training, that was unheard of. So I took some risks [with the barn]." That meant bringing in

The Mid-Ohio Memorial Catalogued Trotting Sale in Mt. Hope, showcasing horses bred for energy, gait, aesthetics, and money, is an annual event that attracts a large Amish crowd. Photo by David McConnell

new bloodlines, breeding Standardbred horses for the form and gait of show horses, with a price tag to match. Suddenly, there was a new interest in speed, endurance, and energy, qualities associated with a race horse.

In the Midwest, where Amish shop culture has created increased economic stratification and where the decline of farming has separated many Amish from their traditional interactions with horses in the fields, good riding and buggy horses are appreciated by well-to-do Amish with money and leisure time.[17] Northern Indiana Amish settlements are said to be the most progressive for high-end horse breeding, with Ohio not far behind. "There are a tremendous amount of breeders out there," an Ohio horseman told us. A veterinarian commented, "The Amish aren't supposed to be proud, but sometimes Amish horses are really well bred. It's like us driving a Corvette. They get a really, really nice horse." Among the Amish, a horse and buggy signals one's economic status in a way that dress never can. An Amish man can spend upwards of ten thousand dollars on a "fancy" buggy horse.[18] But some Amish are critical of their neighbors who own these flamboyant horses, symbolizing worldliness and pride. "We call them the 'Rocky Boot' crowd," said one Amish farmer. "They just have this air about them. Money talks."

Breeding pedigreed lineages is, by default, breeding horses that are candidates for the show ring or for harness racing, activities that run counter to traditional Amish values. Amish breeders invest in a mare with a good pedigree and then mate them with proven stallions to produce foals with desired traits. Most foals will be sold to non-Amish owners as yearlings, but especially promising horses may be sent to a trainer, who then supervises their racing careers. A foal with a superb lineage can bring as much as sixty thousand dollars, although most sell for ten thousand or less. Colts who do well at the track are raced for as long as they are winning and then sold at auction for breeding or to serve as buggy horses, or both. Since Standardbreds are trained to pull a sulky, they make energetic buggy horses. Even with a detour through the racetrack, many Amish Standardbreds come back to the community in harness on a fancy buggy.

Investing in high-end horses has become an outlet for well-to-do Amish who love the competition of the racing world. One Amish breeder described his Amish investors as follows: "They didn't want to have anything to do with the actual horse operations, they didn't want any headaches. They just like to say, 'I own a part of him.'" Amish breeders may own shares in horses that actively race, but the racing itself is conducted by a broker. The Amish generally do not accept money from a horse's racing earnings, but they follow their horses on the racetrack videos. Says one horse owner, you "gotta have something to be proud of." He defends the business, saying, "Horses have an appeal, a cultural element." The cultural relevance of horses for the Amish is undeniable, but high-end breeding, incorporating pride and profit, may be seen as skirting dangerously close to the vices of the outside world.

Amish owners of high-class stallions hold their animals at stud, and stallions with strong pedigrees are in demand. Stud fees for a pedigreed stallion can be in the thousands of dollars, and a stallion can mate with dozens of mares in a season. Making the right genetic choices is the key. Even lower-order Amish are interested. One breeder gives his Swartzentruber neighbors a substantial discount to breed to his stallions. "I just do it for $100," he said. "I come home at midnight, and there's [a guy] sitting out there" with a mare ready to breed. But pedigreed horses have exorbitant price tags, and breeders come under scrutiny from church members. One man commented that if your horse sells at auction for forty thousand dollars, "now everybody knows your income." This new generation of Amish horse breeders are hardworking and passionate, but they push the limits of traditional Amish values. Struggling with his ambivalence, one Old Order man at the forefront of high-end horse breeding mused, "I try not to make the community uncomfortable. In everything, you have people talking negative, but the benefit angle has helped. We didn't just have tunnel vision. I use local businesses, I want to keep the money in the community. Where we're at is not because of me, just the support I got. It's down the road I owe it back."

Modern technology has also made its way into the Amish horse barn. Portable ultrasound machines that evaluate a mare's reproductive status and monitor her pregnancy have been readily adopted by Amish horsemen. One man who taught himself to use his portable ultrasound machine said, "I go to 15–20 places a day, to do ultrasound checks. Even the Swartzentruber are okay with this." Higher-order Amish can use artificial insemination (AI), so breeders collect and ship semen straws to other farms, vastly multiplying the value of a desirable stallion. A strong stallion can produce a dozen or more semen straws every two days, each of which can sell for up to five thousand dollars. The mechanics of collecting, packaging, and mailing time-sensitive semen straws on dry ice to anxious horse owners all over the continent is nerve-wracking. One barn owner described in detail a frantic episode in which the shipping company failed to deliver a semen shipment to a Pennsylvania barn. The mare was due to ovulate the next day, the stallion would not cooperate for semen collection, and the private driver hired to rush the straw to Pennsylvania was waiting on the whole process. On another occasion, "ten [semen] containers got left behind." Recalling that they cost four thousand dollars a straw, he said, "I just about puked." But despite the challenges, artificial insemination combined with live breeding at the holding farm is a major source of earnings for the owner of a sought-after horse.

Tensions between traditional Amish horse culture and high-end horse breeding are evident in the annual event known as Horse Progress Days.[19] For twenty-four years this two-day celebration of Amish horse culture has had as its mission "to encourage and promote the combination of animal power and the latest equipment innovations in an effort to support sustainable small scale farming and land stewardship."[20] Horse Progress Days was created to showcase draft horses from "working farms" that hold to the agrarian ideals of simplicity and frugality. In recent years, however, entrepreneurial breeders have found their way into the program, rankling many traditionalists. One farmer commented that the high-end breeders of draft horses had "weaseled in," implying that what their horses represent compromised these long-held values. These fancy, pedigreed draft teams from monied barns introduce an element of competition into an event designed as a friendly, good-natured exposition of simple values.

Pedigreed horse breeding has become a lucrative and compelling business for some Amish. It has provoked new ways of thinking about investment, leisure, and competition among the Amish, simultaneously reinforcing differences in class, income, and values. The "Current and Fashionable Genetics" advertised by one Amish-owned horse barn contrasts sharply with more traditional values.[21] As Amish horse breeding has moved into the twenty-first century, it has generated challenges to longstanding Amish values that remain to be resolved.

Whitetail Deer Breeding

In October 2014 a whitetail deer harvested at an Amish-owned hunting reserve in Holmes County, Ohio, became the first in the state to test positive for CWD, chronic wasting disease, a neurological disorder found in deer, moose, and elk.[22] Daniel M. Yoder, the Amish owner of World Class Whitetails Hunting Preserve, in Millersburg, Ohio, pleaded guilty to six misdemeanor charges for violations of quarantine and other offenses.[23] Ultimately, 510 captive deer were euthanized, 19 of which tested positive for CWD.[24] The Ohio Department of Agriculture, which oversees deer-farming operations, also reported that two of Yoder's ear-tagged deer were harvested outside his fences, raising the possibility that CWD might have spread into the wild-deer population.[25] As of 2017, state testing in the vicinity had not revealed any spread of the disease to wild whitetails.[26] But some of Yoder's Amish acquaintances chastised him for openly flouting the regulations. "I know the guy," said one man. "Hopefully it won't spread, but that put a hurt on the guys raising the deer" in the area.

Most nonhunters have little idea that raising captive whitetail deer has become a growth industry in the United States over the past twenty years. Even less would they imagine that Amish men in communities from New York to Missouri raise from dozens to hundreds of deer in eight-foot-tall fenced enclosures on their rural properties. In 2012 there were more than four thousand deer farms in the United States, more than six hundred of them located in Ohio.[27] According to one former Amish man, "In the state of Ohio there's definitely a majority of the deer farmers are Amish." Other states with large Amish populations are also well represented in deer farming. Small-scale deer raising is nothing new, but commercial deer breeding gained traction in the 1980s.[28] An Amish hobby farmer in mid-Ohio produced a fawn, Patrick, whose antlers grew abnormally rapidly.[29] As a one-year-old he had a twelve-point rack instead of the five-inch nubs normal for that age.[30] As he grew, his antlers developed into an exotic, "atypical" but sought-after configuration, making him a deer superstar. Patrick was eventually sold to a Texas deer breeder, and trophy deer bred today still trace their lineage back to Patrick. The industrialization of deer farming had begun.[31]

Amish deer farmers do not simply capture wild deer and put them into pens. They purchase animals at auction, selecting those with genetic traits and pedigrees they admire. The doe fawns are kept for breeding or sold as yearlings or bred does. Male fawns are raised and then sold, usually as three-year-olds, to private hunting ranges. Some captive-bred deer are harvested for venison, but an Amish deer farmer noted, "There's no real money in raising deer for meat." The money comes from antlers. If you have been lucky, or savvy, in choosing the lineages you breed, your male fawns will grow impressive racks and can bring high prices as "shooter bucks," hunted on ranches for their heads.[32] And if your stars are aligned, one of your buck fawns will be really special, bringing huge sums as a breeding male. Buck superstars are given catchy names, and they can write an Amish deer farmer's ticket to fame and fortune. Generations of selective breeding have yielded animals that produce amazing racks, which are scored using a series of measurements. While a score of 140 is typical for a fine wild buck, some farmed bucks today have antler scores of 500 or higher. The Boone and Crockett Club, a hunting and sportsman's organization that recognizes trophy animals, doesn't recognize trophy awards from deer harvested in hunting ranges, but the clients who kill these deer go home with a dramatic head mount.[33]

The Top 30 Whitetail Auction, held every year by the North American Deer Farming Association, typically includes several Amish whitetail deer farms among its thirty "top breeders." The auction program includes full-page advertisements showing dramatic head shots of bucks and their best male offspring, testimony to the bucks' ability to pass on desirable traits.[34] The non-Amish owner of a highly competitive deer farm described deer breeding this way: "You're trying to create a perfect buck, and then do it over and over." An Amish wild-deer hunter who vocally opposes deer farming said, "What you have with the breeding is deer that are so bred for antlers that they become unhealthy. They literally have to cut the antlers off of them, because they cannot lift their heads. I call it 'deer porn.' To me, it's not natural. It changes the expectations." An Amish deer breeder, however, had a simple answer to this critique: jealousy. "Because," he said, in the wild-deer population "they can't produce anything like this."

Deer farming can be a comfortable livelihood, but the public-health risks of CWD mean that there is considerable governmental regulation. One man with a long history in the business said, "Thirty years ago, when it started, it was a lot different. That changed in about 2004. Today they test the deer. All of a sudden they had to monitor these deer in order to cross state lines. Some guys decided they wanted no part of it and got out." Amish deer farmers still bristle at the regulations. "We're the most overregulated industry there is," said one. Another said, "It's a business. The DNR [Department of Natural Resources] doesn't like us because they think we've got their deer." The specter of CWD spreading from captive deer herds to wild deer has been a major public-relations challenge for whitetail deer farmers. An Amish deer farmer acknowledged that the Yoder controversy significantly lowered the prices his deer were bringing. "You always knew CWD would hit, but we didn't know how hard it would hit." Captive herds have been implicated in other states as a major vehicle for the spread of CWD.[35] Where CWD has spread to wild deer in some states, its control has been extremely expensive, and its impact on wild hunting and deer consumption could be devastating.[36] But some Amish deer breeders still reject the idea that deer farms are at fault.

In many states a strong lobbying effort, backed by Amish deer breeders, has shifted regulatory oversight of farmed deer from the Department of Wildlife to the Department of Agriculture, where breeders find a more sympathetic bureaucracy. Even under DOA regulation, however, Amish deer farmers must record breeding activity and deer movement among farms. This allows public-health officials to track potential disease spread, but it also helps Amish breeders to document their animals' pedigrees. "Record-keeping is part of the law," an Indiana Amish deer breeder said. "I could think that the regulations are not necessary, but I know why they are in place." But many Amish deer breeders believe the concern over CWD is overstated, and some even subscribe to conspiracy theories, suggesting that CWD was manufactured in the laboratory by Department of Natural Resources scientists in order to destroy the deer-farm industry. These frustrations have promoted the rise of professional advocacy groups, such as NADeFA and state-level deer-farming alliances, who lobby state legislatures for business-friendly policies.

Despite the regulatory challenges, supplying well-antlered bucks to hunting ranges is a lucrative business. A successful Amish deer farmer in the business for more than twenty years told us: "For quite a few years we didn't sell anything for under $12,000 or $15,000. Bucks could go for $65,000. Your price is reflective of what your bucks look like. It's a genetic thing. Deer raising is still probably better than any other lifestyle. I can feed a deer for a whole year for $300." As with horses, artificial insemination is common among Amish deer breeders in the higher-church districts. The easiest way to access the genetics of a distant breeding line is through artificial insemination. If you own the buck, even better. A deer auctioneer told us: "If you have a premium buck that has the genetics that people are looking for, they ship semen all over the United States, and they'll get as many as a hundred and fifty breedings per collection. So they collect a lot of these deer twice a year. If you do the math, if you can sell a hundred straws a year at $5000 a straw, that's a lot of money." Shooter bucks bring a steady income, and stags with truly exceptional antlers can sell for tens to hundreds of thousands of dollars. One auctioneer, a former Amish man who sells these premium bucks all over the United States, said, "The top deer that I've sold so far individually—one deer—was $490,000. And that buck has paid itself off [to its owner]."

Ultraconservative Amish are not typically involved in deer breeding. Some Andy Weaver Amish breed whitetails, but as with horses, they cannot use artificial insemination or sell semen. In one creative workaround, an Andy Weaver breeder sold the semen rights on his high-end buck to a non-Amish man but kept the deer at his Amish farm for live breeding to does. To access genetics from other breeders, Andy Weaver farmers purchase a bred doe that has already been mated to a desirable sire. Conversely, if another breeder wants genetics from an Andy Weaver buck, he has to buy a bred doe—at considerably more cost than a semen straw. Said one Ohio deer farmer, "[A certain Amish breeder] was very famous for doing that. He would breed these does to his famous 'Maxbo Ranger' buck and they were just pretty much $20,000 and up for whatever he would sell. Well, he had that buck for eight years and he sold it and it brought $215,000."

As with horse breeding, the Amish community is quietly divided about deer breeding as an appropriate Amish livelihood. Some see it as an extension of the farming lifestyle, fostering the same work ethic and family values while providing a very good living for a family. Said one Amish observer, "It's an animal, like a cow. Just because it looks like a wild deer, that deer was never wild. It was bred and raised from captivity. I can see how some people would think that they should be wild, but that is a totally different sect of deer." Other Amish, however, are uncomfortable with deer breeding as an Amish enterprise. They note the contradiction inherent in the industry: deer are "livestock" when being bred, but then are "wild animals" when they are released on a hunting preserve. "When [deer breeding] first came out, people just went, 'Oh it's another good way to make a living,' but people are having reservations as time goes on," said a New Order man. When asked how he felt about deer farming, an Amish furniture-store owner commented, "I have a short answer. I do not think it's an alternative Amish life, back to the land and so forth. It's so unnatural, that's not what we're about. It's more related to the money, not doing the right thing to the land."

Many Amish combine a real affection for their animals with a utilitarian perspective on the ultimate market for their deer. But deer breeding is so entangled with the economic system of the non-Amish world that it seems to challenge the "separateness" of the Amish who practice it. It commits them to a business that is in almost every way "worldly," not least of all because it makes some of them very rich. Breeders create animals with exaggerated antler racks that exceed the normal limits of deer in the wild, producing new genetic entities in pursuit of profit and constructing an entirely new relationship with a wild animal. The choices that lead some Amish to practice this profession is a testament to their ability to recalibrate the fine line that separates traditional Amish sensibilities from those of the outside world.

Technology and Politics in the Breeding Barn

Alternative animal breeding is widespread among the Amish, but many non-Amish people also breed show horses, shooter bucks, and designer dogs. In these occupations, more than in almost any other, the Amish interact regularly with their English counterparts, and both produce largely for the non-Amish consumer market. As a result, opportunity for ideas and tools to migrate from the English to the Amish community is especially pronounced and has affected the Amish community in two areas: politics and technology.

In one sense, the Amish are largely apolitical, but like their rural non-Amish counterparts, they don't like regulations, especially those that hurt their bottom line. Amish businessmen are always looking for ways to streamline their operations and maximize their profits. A non-Amish business inspector commented, "The Amish don't like government intervention in any manner. Although they deal with it all the time, they will be the first to pick up a phone and call somebody in government if it benefits them. I have a lot of Amish friends, and they are very astute politically. Anybody who thinks the Amish aren't political is dreaming." Within the lucrative deer- and dog-breeding industries the Amish have taken political activism to a new level. Many Amish breeders belong to state or national advocacy groups with social, educational, and active lobbying functions.[37] Amish may even serve in leadership capacities. An established Amish dog breeder and puppy buyer has been the chairman of the board of the Ohio Professional Dog Breeders Association for several years. A recent newsletter announced the formation of "a new national alliance . . . to help educate state leaders . . . and combat legislation on a State and Federal Level."[38]

As a reason to join, the OPDBA states: "Regulations are coming. . . . We must be prepared. The pet industry must band together."[39] While such occupational activism is familiar to English citizens, lobbying that advances their economic interests is a relatively new political role for members of the Amish community.

Genetic tools and technology also migrate into Amish barns from the English world. Although the Amish have little formal training in science, breeders have appropriated genetic tools like DNA microsatellite and SNP markers (to track deer pedigrees), hormone assays (for detecting estrus in females), and cryogenics and artificial insemination, all in pursuit of breeding outcomes that will sell. Animal breeding also produces offspring with new combinations of genetic traits, which is one of the processes that underpin evolution, a concept the Amish utterly reject.[40] In fact, there is a longstanding debate among the Amish about the use of mules, which were adopted in the Lancaster area in the mid-1800s.[41] The Amish must somehow reconcile a very goal-oriented, science-infused breeding business with their understanding of nature and of their livelihoods. They do this by compartmentalizing the technologies as tools, giving little thought to how these technologies might be used in other settings. Asked if he thought the Amish attached any wider scientific significance to the selective breeding they practiced, one Amish man replied: "It's husbandry. In the terms of the dog market they're going to see [crossing between breeds] as creating a new kind of puppy that will be appealing. In terms of horses or cattle they're going to see, this horse produces amazing offspring. They're going to see it as not genetics, but something in the here and now. But they're not going to see it as science." By compartmentalizing breeding as husbandry, the Amish can thus avoid being drawn into pondering what they consider to be the more suspect implications of science. After all, you don't need to understand the science to make animal breeding work. As one Amish man told us, "We're simple people, chased out of our homeland. We don't make it our business to figure out how things work."

Some Amish, however, do observe the variation in their deer herd or among their exotic pheasants and try to reinterpret the biblical stories in ways consistent with what they see. An Amish pheasant breeder familiar with pheasant species found in Asia noted similarity in color patterns and speculated that they may all have come from one original species. "Look at dogs," he suggested. "They all come from one dog, the original. How can there be so many species? They wouldn't all fit on the Ark." He suggests that while an "original species" might have found a place with Noah, additional variation has been added since the Flood. But, he reiterated, "We believe in creation. When God created it, He said everything was very good. We have a mind to think, we have a right to tame every beast and animal, but we don't want to abuse it." A better statement of the rationale behind Amish breeding industries could hardly be found. Humans take the creations of nature and apply their minds to them. In doing so, they exercise their unique human capacity, and they make animals a source of livelihood. Whether that also includes selecting animals for qualities that appeal to the whims of a worldly consumer population is an issue still open for debate.

III *Reconfiguring Leisure and the Outdoors*

Bringing Nature Home

From Gardening to Herbal and Natural Medicines

From the day they first toddle behind their mother in the bean rows until their hands can no longer hold a trowel, gardening is part of the backdrop of Amish life. Farming as a livelihood has been declining in the face of shop culture, but gardening has largely remained a pillar of Amish home life. While both men and women may be avid gardeners, among the Amish it is central in women's lives and provides them a social network. In our survey of home practices, 98 percent of the 164 Amish families we surveyed reported having a vegetable garden. And every Amish home we visited, except the most conservative Swartzentruber farms, had flower beds surrounding the house or the lawn trees.

An Andy Weaver man assured us, "The gardening is not really a hobby, but part of our way of life." Gardening is a homemaking activity, one that provisions the family and organizes and beautifies the home space. Mark Bhatti and Andrew Church suggest that home gardens act as "everyday spaces [providing] important sites for lay knowledges of, and connections to, nature."[1] Furthermore, they are places of "real and imagined communication and social interaction with family, friends, and other agents."[2] The gardener chooses, handles, and nurtures the plants she wants, which reflect her ideas of beauty, purity, efficacy, and comfort. Because "gardens are places where nature and culture meet,"[3] we must see them as more than just "nature." Despite their constructed qualities, however, gardens provide their keepers with spontaneous opportunities to observe and learn. Gardens can be a window through which we observe Amish interactions with nature.

Amish gardeners often report that they learned their skills and plant lore from their own parents or grandparents, making their garden a historical link to their family and their past. Amish gardens contain the traditional annuals and perennials of northern temperate latitudes, as well as plants for teas, poultices, tinctures, and folk remedies. Medicinal plants, like nettle and jewelweed, are gathered from forests and fencerows. A kitchen garden, containing potherbs, remedies, and other useful species, is a common feature of traditional communities around the world and harks back to the earliest farming experiences of the Amish community in Europe.[4]

For centuries, plants were the most readily available source of active chemicals in the human pharmacopeia. Plants are also appealing as medicines because they are completely natural: raw, unprocessed, and harvested under known conditions. That the plants they need for staying healthy are found growing where they live is logical for the Amish, who believe in a world where God provides for all people. In this chapter, we explore gardening, garden design, and the use of herbal and natural medicines among the Amish, with an eye toward discerning the diverse ways in which the Amish define and use the term "natural."

Gardening for the Table

The Amish are vegetable-garden champions. As noted in chapter 3, Amish gardens provide fresh produce in season and canned or preserved vegetables for the winter. A vegetable garden is a family affair. The husband tills or turns the garden in the spring, then the wife often takes over, making selections from seed catalogs and planning the plot. Although Amish gardens are large, they usually aren't large enough to accommodate a team of draft horses. More progressive Old and New Order Amish use gasoline-powered rototillers or garden tractors to turn the soil. A New Order minister described how their Ordnung forbids tractor farming, but "we property owners kind of escape the 'no tractor farming' rule because we're not farmers." A New Order woman said her husband does "custom tilling" in spring and fall for the Andy Weaver Amish, who cannot use tillers. "They want it worked up really good," she laughed. While some men may participate in the gardening through the season, it is largely seen as a woman's job. By tending, harvesting, and preserving garden produce, the wife makes a substantial contribution to the home economy, saving money and producing healthy, fresh meals. Gardening also supplies chores for children, helping to instill in them responsibility and a strong work ethic, virtues that many Amish lament have been lost since the spread of shop culture. The closeness of the vegetable garden to the house makes it a natural extension of the woman's housekeeping realm, a managed space under her control that provides the family with comfort and sustenance.

Gardening is also biblical. There are dozens of references in the Bible to gardens and the fruits and flowers, herbs and spices they yield. They are refuges, sources of nourishment, and places of water. The familiarity of gardening, as well as the challenges and rewards it entails, makes it a frequently used metaphor in church sermons. A New Order minister explained, "The nature illustrations in preaching are used a lot, anything from trees to gardening. Your preaching illustrations come out of everyday-ness and so if you garden, there's illustrations. So church purity might be metaphored to a garden with weeds, and how it takes effort to keep the weeds out, how the young plants must be protected." Gardens can also carry moral connotations. God created a garden for Adam and Eve, which had "every tree that is pleasant to the sight, and good for food; the tree of life also in the midst of the garden, and the tree of knowledge of good and evil."[5] Disobedience can result in being cast out of the garden, or having your vineyards and fig trees devoured by the palmerworm.[6] A young Amish woman who obviously had hands-on experience recounted weeding her garden. "Sometimes I get so annoyed at Eve," she sputtered. In contrast, an Old Order furniture maker mused, "The Lord is a partner in the garden." Among simple, godly, and appropriate activities, especially for women and children, gardening is at the top of the list.

"Nearer God's Heart in a Garden"

"Nearer God's heart in a garden / than anywhere else on earth." So goes a well-known couplet that appears on benches and sundials in gardens all over the English-speaking world.[7] While vegetable gardens are utilitarian, flower gardens are beautiful. For Amish and non-Amish alike, flowers are reminders of the beauty of the world. Flower gardens offer Amish women and men a creative outlet, with nature itself as the palette. They are places where one can feel inspired, nurtured, and safe. For the gardener, they are personal spaces where one can nurture plants and objects that realize one's notion of beauty. An Old Order woman recollected, "My grandma was a serious gardener, and she had a huge garden, way more than she needed. My dad also loved gardening. Grandma was a weeder, she taught that to me. She had roses, the nicest roses, and mums, peonies. When she was older, we helped her spade her garden. She taught me about how to redo the mums. All our children had their own gardens." While most Amish gardens are not formal, they are certainly planned and bring together plants that are beautiful, useful, and perhaps even meaningful. An older Andy Weaver woman took us into her yard and showed us the small area she had created as a memory garden for her husband, who had died the previous year. Located in front of a white picket fence, it contained only species whose flowers were pure white.

Like English gardeners, the Amish use commercially available annuals and perennials, some started from seed but most purchased at Amish-owned greenhouses. Small, family-owned greenhouses are an obvious choice as a safe and viable off-farm business that allows for an ongoing connection with nature and a way for all family members to work together. "I go to a local greenhouse place," an Old Order woman told us. "I've thought about saving my seeds, but I'm not that interested."

These young greenhouse workers decorated their bonnets with the extra flowers they trimmed from the plant benches. Photo by Marilyn Loveless

A New Order woman said, "We do a lot for cut flowers. We have perennials and some annuals. . . . We do sunflowers, zinnias, geraniums, impatiens. Also lilies, Shasta daisies, black-eyed Susan." But only a few of the women we spoke with were specifically interested in planting native perennials, and none mentioned efforts to foster pollinator diversity or implement ecological gardening principles.

Almost all the Amish we met could identify the common tree species of their landscape. Furthermore, most were familiar with the common, often introduced weedy plants that grow along roadsides and in old fields, and they used European species like mullein, clover, burdock, and mint in folk remedies. But relatively few Amish were dedicated botanists, exploring pristine habitats for uncommon native species. The best wild botanists were butterfly watchers, who were attentive to nectar plants and to the wild plants used by caterpillars as food. One Andy Weaver man, a birder and butterfly aficionado, was exceptionally interested in native species. When he found a native wild orchid growing near his home, he lovingly surrounded it with a small fence and a sign urging its protection.

Among the progressive Amish, some pursue their passion for gardens as collectors, almost akin to keeping exotic pets. An Old Order man whose hobby was hostas belonged to two circle-letter groups specializing in the plants; contributors shared tips on propagation, disease control, and specialty species. Among the conservative Swartzentruber Amish, vegetable gardens are a given, but flowers are seldom used as a landscaping tool. One former Swartzentruber man told us, "We were not allowed to have flowers." A man from the Geauga settlement agreed. "We were always told that the beauty of nature is . . . what's the right word? To dress up our front yards with flowers was prideful." The man's wife recounted, "My mom liked flowers. But she had her flowers planted in the tea bed, where the tea was grown. I remember, having church at our house, she'd actually pull all of the flowers off so the people don't have to see the flowers."

The Public Landscape

The more liberal Amish are diligent and creative in fashioning well-manicured, inviting landscapes around their homes. While gardens can be very personal, they are also public. Landscaping is a way of setting yourself apart and advertising your social niche within the Amish community. A neatly manicured lawn tells the world, "This family values order." A firepit, a play yard, a gazebo, or a clothesline says much about how you fill your time. Many Amish spend considerable money and effort on the curb appeal of their homes. In St. Ignatius, Montana, New Order homes have carefully landscaped yards, expansive lawns, landscaping fences, lawn-furniture sets, and sunken fireplaces, in stark contrast to the homes of some of their poorer, non-Amish, sometimes tribal neighbors. The same is true in Lancaster, Pennsylvania, and other settlements throughout the Midwest. When we arrived at one Old Order home, a daughter was on her hands and knees trimming the grass along the sidewalk with a pair of scissors. At another home, we complimented the owner on his lovely yard. Beaming, he told us, "The realtor said, 'Just drive to the nicest place you ever saw, and that's our house.'"

Much has been written about the ethnoecology of the American lawn, and most Amish share with their English neighbors a commitment to a manicured home landscape. While a Swartzentruber farm will devote only minimal space to a yard, lawns are common among more liberal Amish. Asked if there was a premium on having a neat, clean lawn free of weeds, an Old Order Ohio man replied, "Yes. There's a lot of Amish people in our neighborhood whose lawns are largely mixed with weeds, and it doesn't bother them at all. It bothers me. And there's a lot of Amish people of our kind that will get lawn care companies that will come in and spray, fertilize, and so on. I like to look out and see grass without plantain or dandelions." Despite

Keeping a neat, manicured lawn, often with intricate landscaping, is important to many Amish families in the more progressive affiliations. Photo by Marilyn Loveless

its cultural significance, "the lawn is by no means an indigenous ecosystem and, as a result, the requirements for its propagation are high."[8] In pursuit of a pristine landscape, the Amish use a variety of yard chemicals, including Round Up, 2,4-D, and Sevin. Paul Yoder, editor of the Mennonite magazine *Pilgrim's Pathway*, observed that a century ago Plain people's homes typically had a side yard full of fruit trees. "Slowly," he laments, "the interest seemed to fade out." He continued, "We are influenced by our vain society. Extensive landscaping, groomed lawns, and non-productive, sterile accent plantings have replaced productive plantings. Priorities change when such a pattern becomes established. People decide they don't have space for fruit trees, yet spend hours pacing behind lawn mowers and mulching ornamental shrubbery. Values need to be discerned when this pattern sets in. We may not all need an orchard, but we must not become obsessed with vanity."[9] A New Order man expostulated: "All those big yards! And the lawnmowers get bigger and bigger, while the gardens get smaller."

When homeowners outsource their landscaping to a lawn-care company, they in a sense cede their connection to nature, but every real garden carries the fingerprints of its owner. Gardens reflect priorities, aesthetics, and values, and like owning and nurturing animals, gardening teaches those who tend them about how life renews itself. Plant materials appeal not only to the eye but to the senses of smell, taste, and touch as well. Because plants are a source of nourishment for people and for animals, they are wholesome, contributing to growth and health. Growing and using plants to supplement human health forges an additional connection between nature and people. For the Amish, gardening embodies God's promise to provide for human needs.

The Landscape of Natural Healing

In 2013 the case of Sarah Hershberger, a ten-year-old Swartzentruber Amish girl diagnosed with an aggressive form of lymphoma, made national news when her parents decided to stop chemotherapy treatments at Akron Children's Hospital in Ohio because they were inflicting too much pain on their daughter. The hospital went to court to assume medical guardianship of the girl to ensure that she would finish the treatment regime. The Medina County Court ruled in favor of the parents, but the Ninth District Court of Appeals reversed that decision, prompting the family to flee to Mexico with their daughter days before a court-appointed guardian was assigned. Libertarians and other outsiders sympathetic to the parents publicly pilloried the hospital's conduct and the court's decision, while considerable informal pressure was brought to bear on the hospital by Amish themselves. Facing a public-relations nightmare, the hospital dropped its suit, and the Hershbergers returned home, where Sarah was pronounced cancer free and in good health several months later.

Less noted in the heat of the legal battle, however, was that the hospital initially went to court "after the family decided to treat Sarah with natural medicines, such as herbs and vitamins."[10] A relative of Sarah's who was involved in the decision noted that "we as adults were not doing all the things the doctors wanted us to do." But he acknowledged that the family's decision to stop treatments was informed by his own experience with other such patients. "Here's what we discovered in lymphomas," he told us. "Most lymphomas respond very well to chemotherapy. But even before they [the doctors] say it's done, it's high time to get that person's immune system back up and into nutritional therapy. And we are very confident in that." The Hershberger family's response, it turns out, was not a simple decision based only on their reading of God's will, as was widely reported in the media. Rather, it was a calculated choice made on the advice of Amish folk healers, whose experience and understanding of medicine were reassuring to them. The Hershberger case raises important questions about the meaning of natural remedies and how Amish folk healers and consumers of natural health-care products appeal to religion and science to justify their choices.

The pluralistic nature of Amish health care has been well established in the scholarly literature. When Amish get sick, they often draw on combinations of folk, alternative, and mainstream medicine.[11] Studies of the progressive settlements in Lancaster and Holmes Counties, for example, found that 90 percent or more of Amish respondents reported seeing both chiropractors and medical doctors, while 80 percent used dietary supplements and about half used prayer for healing, as well as cleansers and detoxes.[12] In general, individuals from conservative Amish groups are less likely to access the modern health-care system than are members of more liberal affiliations.[13] "We try to avoid going to the doctor if at all possible," commented one Swartzentruber mother of ten. Less explored, however, are the meanings Amish attach to "natural" remedies and the model of bodily healing that underlies them. Some families still rely on natural remedies as the first line of defense, while others turn to them later on in the course of an unresolved illness.

Outsiders are sometimes perplexed by the Amish insistence on mixing folk and alternative remedies with biomedical options. On the one hand, the Amish are held up as a repository of age-old wisdom about herbal teas, tinctures, and ointments and as an example of how to combat the "unnatural" sterility, high cost, and impersonal character of biomedical care.[14] According to those who hold this view, the Amish have bravely resisted the worst elements of the modern health-care system, electing for a more natural approach, such as setting up their own birthing centers or choosing to die on their own terms in their homes. Their lifestyle, especially physical activity and interaction with animals and pathogens, also brings health benefits, such as the low levels of asthma reported in a 2016 study published in the *New England Journal of Medicine*.[15] A more critical view, however, sees the Amish as engaging in a number of unhealthy practices, including a diet heavy in carbohydrates and fats, pipe and cigar smoking in the conservative groups, lack of regular preventive health care, and relatively low immunization rates. The Amish are also portrayed, even in some cases by themselves, as gullible consumers whose scientific naïveté leads them to accept quack medicine and far-fetched cure-alls. In worst-case scenarios, as in the Hershberger case, they are accused of turning their backs on the clear medical consensus in favor of unproven "natural" approaches and an appeal to faith.

Amish conceptions of what is natural shape their health-care choices. There has been an important shift in Amish practice, from using homegrown or locally collected herbal remedies to consuming commercially produced "natural" medicines at home or in clinics run by non-Amish or Amish folk healers. The use of packaged vitamin and mineral supplements, tonics, dried herbs, and assorted pills has soared among the Amish because they are less expensive than conventional medicines and do not come with the long list of side effects that pharmaceuticals carry. Home-administered remedies allow for more patient participation than does physician care. Amish consumers of these products justify their choices by mingling the usually competing discourses of religion and science in creative and sometimes incongruous ways. Their choices of "natural" medicines often reflect a complicated mix of family traditions, testimonials from those they feel they can trust, their interpretations of the science surrounding a given illness, and affiliation-based identities.[16]

Herbal Remedies and Natural Healing

It is risky to make generalizations about Amish views on health and the human body, since such understandings are highly idiosyncratic and depend on personal history and experience. Most Amish, however, express at least a little skepticism about modern conventional medicine, which one medical anthropologist describes as based on "the hope-filled notion that through technological advances we will ultimately transcend all limitations seemingly placed on our bodies by biology and nature."[17] The roots of this skepticism lie in a conception of the human body that is aptly described in Kraybill, Johnson-Weiner, and Nolt's 2014 book: "The Amish inhabit a sacred world filled with the spirit of God, who intervenes to bring about certain outcomes, and Satan, who seeks to distort God's plans. They see nature as God's handiwork and think that the more one embraces nature, the closer one walks with God. Likewise, because the body is a natural organism, the more one treats its ills with natural remedies, the more one is in tune with the mysteries of God's intents."[18] This description calls attention to the body less as a locus of individual expression than as a place where individual autonomy is mediated and restrained by the church community in a way that is consistent with divine will. The driving logic behind this view, as one Amish folk healer explained, is that "God did not make very many things for no reason. That philosophy goes a long way in healing a person that has a chronic illness."

Herbal remedies made from homegrown or local plants are a straightforward extension of this view of the body as best cared for with naturally available resources. Almost every Amish home has a copy of Rachel Weaver's *Be Your Own Doctor*, one man told us.[19] "It's the number one book in the Amish community." One former Amish woman from the Geauga settlement remembered that tea made from nettles and alfalfa was popular for morning sickness for pregnant women.[20] A Swartzentruber bishop told us, "The lowly dandelion has some very powerful attributes. I've seen people with a handful of warts, and one pound of dandelion leaf powder takes care of them. They drink it in a liquid mixture. I call it the 'poor man's medicine.'" A Swartzentruber mother treats infections and boils with a poultice made from two quarts of milk, a slice of bread, and the roots of a wild plant. "You want it fairly warm, and it draws the pain out so hard." She also told us that "vinegar is good for just about anything," and "medium red pepper, not mild, we go through pounds of it." On one occasion, we visited a home where the women were in the midst of making a year's supply of "super tonic," concocted of garlic, horseradish root, onions, and ginger, all cured in vinegar for six months and then strained. The woman of the house was effusive in telling us how a teaspoon per person, passed around the dinner table every night, had kept her family healthy through the winter.

Perhaps the most well known Amish folk remedy is B&W (Burn and Wound) Ointment, a treatment that incorporates burdock leaves. Burns are a particular threat to the Amish community because heating, lighting, and cooking often involve an open flame and flammable liquid fuels. As he describes in *Comfort for the Burned and Wounded*, John Keim discovered the healing qualities of burdock while standing, distraught, in a field shortly after his child was badly burned. He felt that the Lord intervened, and he had a vision that he could use burdock to heal his daughter. B&W Ointment, a mixture of honey, lanolin, olive oil, aloe vera gel, comfrey root, and other ingredients that is applied with a dressing of burdock leaves, is widely used and praised throughout the Amish community. Though it has not undergone clinical trials, several studies have affirmed its efficacy for first- and second-degree burns.[21] "We have people among us who have been trained," said one Andy Weaver man, "because you have to know the role of B&W. People with third degree burns will dehydrate and need an IV." We also talked with several Amish individuals who had successfully used steamed burdock leaves to wrap and treat wounds from severe cuts that initially required emergency-room visits.

Folk remedies are popular among the Amish because they use ingredients that seem close to nature, are far less expensive than medical treatment, and rely on the experiences of familiar adults, especially women, whose wisdom has been passed down through the kinship system. Many Amish women include their regularly used herbal remedies in their garden plan. Others collect material from the wild. About one-quarter of respondents to our survey said they collected common wild plants for herbal remedies, with burdock leading the way, followed by peppermint, plantain, comfrey, yarrow, nettle, chickweed, jewelweed, and elderberry.[22] One Andy Weaver man who recorded all his experiences with herbal remedies in a personal notebook said, "I've got a mix that eases headaches. It's got tea tree oil, peppermint for pain relief, and marjoram. I had to really rub it in for migraines." In some cases, Amish families use herbs as a platform for a home business, growing and selling plants and plant-based tinctures and remedies from a shop adjoining their home.

One Amish healing technique, powwowing, is a "traditional magico-religious practice" guided by an expert practitioner, or *braucher*, that uses "words, charms, amulets, and physical manifestations" to heal people or animals.[23] David Kriebel's definitive study reveals that powwowers "have often practiced herbalism as part of their healing activities (brewing teas, formulating 'vitamins' and so forth)" and notes that some of these concoctions worked but others were valueless and even harmful.[24] Powwowing is still practiced, usually in secret, in only a very small minority of mostly ultraconservative churches, and it is widely dismissed as "witchcraft" by members of liberal affiliations. But the core assumption underlying powwowing—that faith in God heals illness—can sometimes be seen in the practice of "divine healing" in more liberal Amish circles, more as a last resort than as the first line of attack.[25]

Nature in a Capsule: The Commercial Connection

Over the past few decades, commercial natural-medicine products have become increasingly popular in Amish circles, with some families spending hundreds of dollars a year on supplements. Seventy-seven percent of respondents to a 2015 survey in the Lancaster settlement reported using herbal or vitamin and mineral supplements, while 22 percent reported using doctor-prescribed medications. Of those who used supplements, women used greater quantities, with more frequency, than men.[26] Distinct from traditional herbal remedies, which are homemade from the garden or from locally purchased ingredients, these products are commercially fabricated and mass marketed, even from sites overseas. Some Amish buy their supplements and other packaged remedies at their local bulk-foods store or from a known "Amish doctor"; others order them and have them shipped directly to their homes. Like the consumption of herbal remedies, the consumption of natural medicines relies heavily on testimonials from trusted authority figures, often relatives or church members, rather than from distant physicians, and it allows patients to take an active role in orchestrating their treatment. Self-medication through supplements and tonics also relies heavily on several "logics of naturalness" that are particularly compelling from the standpoint of an Amish theology of the body.

The idea expressed by Hippocrates—"Let food be thy medicine and medicine be thy food"—has particular resonance for Amish who are trying to fight an illness. To this historically self-sufficient people who see themselves as special in God's eyes, it makes sense that God would provide the means to cure any illness without having to turn to doctors. What you put in your body in the first place is important, and the more natural, the better. "Just raw is as natural as we can get," said an Andy Weaver folk healer who recommended staying away from highly processed foods. "There's no nutritional value left [in store-bought food]," one Old Order man told us. "You have to eat four apples, now, to get the nutrition you used to get from one." A Swartzentruber folk healer echoed his assessment:

> I don't know if you've heard the term "epigenetics." In our community, we have a lot of Amish who have a very poor perception of health and how it is gained or lost. Here's what's happened. Our modern world now is everywhere we are, and cheap food—that is not of much value—is the problem. There is no disease that does not have a nutritional component to fixing it. Some products may not really seem like nutritional support, but their molecular structure is what your body is craving. For instance, for specific amino acids. There's even a name for these compounds. They call them nutraceuticals.[27]

Amish promoters of natural medicines thus see supplements as a way to address an underlying dietary deficiency that is causing illness. The cure is less focused on eating well or changing one's diet than on taking the right mix of nutritional supplements.

Another salient idea that drives consumption of natural medicines is the notion that the body needs cleansing or purging because it has become clogged. One full-page advertisement for colon cleansers in an Amish-run periodical claimed that the number one problem in America is "NOT going to the bathroom" and that "most colons are like a toxic cesspool." Jim's #1 colon pills were recommended to "push it out," while a dose of the #2 pills "scrubs and cleans."[28] Among the ailments these pills purport to cure are gastritis, Crohn's disease, headaches, polyps, and breast cancer. Such products and claims are ubiquitous in Amish stores and in Amish-generated print material across the country. The parallels between the embrace of cleansers in pill and liquid form and the Christian experience of confession, forgiveness, and the cleansing of sins may seem obvious, but they take on added meaning for the Amish. Modernity has, in their view, produced a whole host of polluting conditions to which natural medicines can offer a corporeal antidote. Notions of purity and pollution are also a part of allopathic medicine, of course, but in a very different way. Germ theory rests on the fact that microorganisms, such as bacteria, cause disease,

Commercial natural-medicine products are very popular with Amish vendors and visitors to the annual Tri-County Health Expo in Kidron, Ohio. Photo by Marilyn Loveless

but these pathogens must be exposed to scientifically established treatments, such as antibiotics, in order for purity to be restored.

Amish folk healers and promoters of natural medicines sometimes rely on the homeopathic principle that "like cures like," as well as on an approach known in medieval times as the doctrine of signatures. The logic of the latter is that God made plants in forms that suggest the remedies for which they are useful.[29] "I really do think this is the case," said one promoter of nutritional supplements. "A reputable company I use, they have an extract from kidney beans that is intended to address kidney problems." Acknowledging the dangers of some applications of this principle, he added, "I'd also want to test it to be sure it was efficacious." One folk healer took the principle of likeness, mingled it with elements of homeopathy, and created his own version of a measles vaccine. Based on the assertion that the measles virus emits a characteristic frequency of a certain type of wave, he took tinctures of St John's wort and echinacea, put them in a machine, and tuned them to the same frequency as the measles virus. He then offered them to his patients as a "natural substitute" for the measles vaccine during the 2016 measles outbreak in Ohio, much to the concern of local health officials.

In recent years several Amish-run health-care clinics dedicated to naturopathic medicine have emerged in midwestern settlements. They are legally organized as members-only institutions and therefore can offer alternative therapies that are not sanctioned for use by licensed medical professionals. These clinics often have licensed, non-Amish physicians or nurses nominally on their boards, but day-to-day care and treatment of patients is largely done by unlicensed Amish "doctors" and "nurses" who have a reputation as healers in their community. Noted one non-Amish physician, "They are pushed by an eclectic group of non-Amish from Berkeley liberals to West Virginia conservatives who are trying to bring validity to their practice of homeopathic medicine." These clinics often rely on blood screening and saliva testing to identify how a patient must balance their blood chemistry with nutritional supplements. Some use vague procedures such as brain massage and chelation therapy or employ hyperbaric oxygen chambers in ways that are broader than their use in conventional medicine. The Amish healers who run these clinics fervently believe they are doing a service to the community, and some even offer accommodations to out-of-state patients at an affordable rate. Contrasting his operation to mere reliance on home remedies and mail-order supplements, one clinic director emphasized the importance of counseling his patients in a Christian environment. "We're on the next level—integrative medicine," he told us. Such folk healers are usually trusted because they are assumed to be working in a Christian framework for the good of the patient and can attend to the social and the spiritual dimensions of healing.

Using the Language of Science to Validate Natural Medicines

Although the lack of science in the Amish school curriculum means that most Amish are unfamiliar with the scientific method, they do understand the legitimacy that science confers in certain areas of human life, including health care. Amish promoters and consumers of natural medicines selectively appropriate the language of science and apply it liberally, and sometimes incorrectly, to confer legitimacy on the use of supplements. Asked if he used science in his approach, a Swartzentruber folk healer who ran a natural-supplement business out of his basement replied, "Very much. I call it 'True Science.' The way I look at medical science, it is horrible. The one main thing that drives medical science is money and the power to keep it there. But to me, science is finding the best way to solve a problem. As far as experimenting, I cannot say that I do that. I read about something, and that's what alerts me to possible treatments. But I refer to experience." This healer expressed serious reservations about pharmaceuticals that "are molecularly changed from their natural structures." He described them as unnatural because "they manipulate organs and bodies" and concluded, "This is not what we want to do. I have things that I'm confident are better than antibiotics." Many Amish herbal-medicine vendors make a similar point. They argue that despite their identical molecular structure, compounds extracted from natural sources (usually plants) are more efficacious than the same chemical compounds synthesized in a laboratory. The difference is that extracts are more "natural" and thus more appropriate for use in addressing human health issues.

For Amish in more liberal affiliations, even cutting-edge medical technologies can be interpreted and used in ways that support Amish views of naturalness. One successful Old Order businessman became intrigued by claims that DNA testing could provide a "readout" of the metabolic and nutritional flaws in one's genome. He subsequently paid tens of thousands of dollars for DNA testing for dozens of his employees so that they could find out which allergies and illnesses were tied to their genetic makeup. "It's like splitting open a tree and seeing where the knots are," he told us. "There's a lot of genetic crossing and this will allow offspring not to have the same weaknesses as the parents." In this case, he relied on xRMD, a company founded by an ophthalmologist, George Rozakis, that advertises the ability to help patients make lifestyle choices that are fully personalized based on measurements of "specific aspects of your genetics and biochemistry." Though xRMD has been criticized for "bio-hype" that is not backed up with published scientific evidence, it appealed to this man and some of his employees because of the customized nutritional recommendations it provided.[30] "You can offset your genetic tendency," he claimed, "by taking herbs and pass on the health benefits to your children."

The love affair one segment of the Amish community has with nutritional supplements, however, has generated sharp criticism from another. "There's people in our community that I think are going overboard in the natural thing," commented one Old Order businessman. "Because if it's called natural, it's okay. But the word natural is very, very broad. The Food and Drug Administration has not defined it. Natural can be anything." Out of curiosity, he combed through a recent issue of the *Budget* and discovered "thirty-some ads for some kind of cure-all." Laughing in disbelief, he reflected, "[The Amish] fall for that more." One Old Order skeptic of pill pushers told us that he had taken some supplements advertised for combatting heart disease to the local pharmacist, who told him they contained a laxative.

For these Amish, usually from more liberal affiliations, the fruits of medical science are perfectly acceptable, even embraced, as long as they are kept in their proper place. Asked why some Amish accept the science underlying germ theory but not natural selection and speciation, a New Order man exclaimed, "Science is not medicine! Science is evolution and half-naked people who came from monkeys." To our surprise, one Old Order woman noted the contradiction inherent in this tendency to compartmentalize the two. "There are a lot of Amish who put their faith in science," she said. "They'll say, 'Oh, homeopathic medicine is witchcraft.' But you can't turn around and then say scientists who believe in evolution are monkey-lovers!" Her comment, though highly atypical, calls attention to the fact that scientific claims are sometimes arbitrarily applied in the Amish community. One Amish man, for instance, argued that the Amish religious view of the world is indeed supported by science. "We definitely don't buy the millions and millions of years," he said, "because we believe we have scientific evidence that is biblical, and that fits with our understanding of the world." A non-Amish physician who works closely with Amish children and parents summed up the incongruous claims this way: "They don't have a standard that's scientific for what evidence is." As a result, it becomes one person's view of what counts as science against another's.

The complex interplay between religious and scientific interpretations was illustrated by a New Order man who asserted that you should never use your right hand to cover your throat if you have a cough. This is because in the earth's magnetic field the right hand is positive and "pulls," whereas the left hand is negative and "pushes." He said that he could cure his son's stomachaches by putting his right hand on his son's stomach and his left hand on his back. "And it honestly does work that way." But he also noted, "Some [Amish] would say, 'That is a hoax. That comes from evil.' " This same individual, however, described as "black magic" the practice whereby Amish herbalists and vitamin peddlers hold a nail clipper on a string and let it swing, pendulumlike, above your hand to "see how many pills you need." Asked why he evaluated the two practices differently, he acknowledged that there was a fine line between the two but elaborated, "Because of spiritual connection. And I'm a firm believer, I really am. The spirit of God, the Bible says, the spirit within you is more powerful than the spirit in the world or Satan's power, when you really have His spirit within you. And so to me, I had no qualms at all that that spirit had any power over me." Pressed further about how he knew that the nail clipper embodied Satan's spirit and not the more benevolent magnetic energy in the left-hand, right-hand example, he remarked: "The nail clipper has absolutely no knowledge of the pills in your hand. To me, there is no *scientific* connection there at all" (italics ours).

In this case, plausibility based on personal observations and reasoning served as his criterion for evaluating whether a claim was scientific or not.

Finding Common Ground

Even though Amish and non-Amish patients look to both medical doctors and folk healers for relief of bodily symptoms, their frameworks of healing sometimes do not align with each other. In some cases, Amish decisions to pursue natural healing approaches do not attract the attention of medical professionals. The decision to purchase a Kangen water ionizer, hooked up to the faucet to separate water into acidic and alkaline streams, one for drinking and one for external uses, is purely private—even if, as one physician noted, "the importance of pentavalent or ionized water can't be found in peer-reviewed journals." In other cases, the perception gap between Amish and medical professionals can be partially bridged through conversation during clinic or hospital visits. A medical doctor who works closely with Amish children with heritable disorders reflected: "There's a disconnect between my view of what's natural and theirs. For example, with metabolic disorders, how could you find a disorder that's any more natural to treat?" However, instead of restricting amino acids in the diet, which prevents the buildup of toxic incomplete metabolic products, "they want to try other kinds of natural remedies that they get at bulk food stores, or goat's milk. It's an irony. So I tell them, 'No, that's a different type of cleanser than what your body needs for this.'"

When the Amish access the health-care system late in the course of an illness or abandon doctor-prescribed or public-health protocols to embrace low-cost herbal or natural approaches, however, their decisions may lead to bewilderment and even outright condemnation from non-Amish physicians. One medical professional related how an Amish patient stopped follow-up protocols after open-heart surgery that cost hundreds of thousands of dollars, choosing instead to use magnets and red yeast rice to lower cholesterol. "It drives specialists crazy," he said. Non-Amish physicians we spoke with were also quite critical of the growing tendency for untrained Amish folk healers to analyze their patients' blood work and make lifestyle recommendations. "They're diagnosing Lyme disease," cautioned a non-Amish physician. "The patients will say, 'I saw the spirochetes in my blood! I have Lyme!'" He noted that such clinics will say they rely on science even when making diagnoses that are at odds with the scientific consensus. "They will pull out their studies, but they can't be replicated."

Some of the naturopathic approaches embraced by Amish healers and consumers can be easily dismissed as unscientific, even "quack," medicine. One non-Amish doctor described the mind-set behind the embrace of the more dubious cures as "suspicion of conspiracy, suspicion of authority, fear that the government is out to get you, low education, and hyper-religiosity." He pointed out that this combination is a "recipe for swindlers" because they can say they've "learned the secrets that others don't want you to know about." He continued: "It's hard not to come across as elitist or ivory-tower-ish when you say, 'Guys, this is not how science works!' The Amish will say [scientists] can't make up our minds, but science is a process. It takes years to learn the scientific method, and there are clear recommendations that no one disagrees with." Since germ theory and the scientific method usually are not taught until high school, even Amish children in public schools miss out on important pieces of the logic of science that underpin the modern health-care system. It may be more difficult, then, for scientifically naïve populations to grasp the causes and long-term health consequences of what Rob Nixon calls "environmental slow violence," such as the entry of chemicals into our water and food supplies.[31]

In other respects, though, the desire of many Amish for low-cost, patient-centered health care with a trusted provider who is open to more natural approaches cannot be so easily dismissed. Who among us hasn't tried to second-guess a doctor's prescription, or fallen prey to the enticement of an unproven treatment for an ailment that hasn't responded to conventional medicine? Amish and non-Amish patients alike often struggle to understand scientific explanations for illnesses and get equally upset when physicians talk over them. We all work to discover natural remedies that make sense for our respective world-views and that might complement or even alleviate the need for an expensive medical procedure or intervention. And if human biology is deeply affected by the environment, as the emerging field of epigenetics shows, it is reasonable to worry about the harmful effects of growing up in the presence of pervasive, human-produced toxins. The Amish may be better positioned than many non-Amish to keep modern medicine in perspective. The medical ethicist Daniel Callahan argues for a sustainable medicine to improve the quality of health "within a finite life cycle in a way that seeks to avoid straining the biological or social or economic capacity of humans to adapt to the progress achieved."[32] The Amish are sometimes able to put the brakes on expensive medical treatments because they have a theological perspective and a support system to fall back on. The Amish enthusiasm for herbal and natural medicines, which medical professionals at times find frustrating, can also be interpreted as an attempt to retain some semblance of cultural autonomy in an arena of life that has become increasingly technological and impersonal.

Fin, Fur, and Feather

Nature-Based Recreation

When the Cleveland Cavaliers upset the Golden State Warriors in game 7 of the National Basketball Association Finals in June 2016, the owner of Lem's Pizza in Fredericksburg, Ohio, posted an unlikely photo on his Facebook page. It showed dozens of Amish teenagers at a "watch party" set up exclusively for them at his local pizza joint. "One of our weak points is that our teenagers have too much free time," lamented an Andy Weaver father. "Sports really consumes the boys." Less than two months later, Amish teens were in the local news again: seventy-three had been arrested for underage drinking at a party in Holmes County that attracted more than one thousand Amish youth from several states.[1] As the psychologist Richard Stevick points out, the media have sensationalized rumspringa and perpetuated numerous myths about this stage of Amish adolescent life.[2] Nevertheless, disagreements over how to control the youth have been at the heart of several major church divisions, and intergenerational conflict remains a central social dynamic in most settlements. In the larger settlements, where youth can choose to join one of more than a dozen informal peer groups of varying reputations, some parents and church leaders have taken steps to increase adult supervision at youth events. For all parents, however, concern about the activities their children become involved in begins well before they turn sixteen and "go with the *youngie*."

Ironically, the successful economic transition of many Amish heads of household from the farm to the shop has created an entirely new social challenge: how to be effective parents for children and teenagers who have a lot of free time and spending money. When a family is running a farm 365 days a year, there is no shortage of chores for children of any age; however, when fathers are working away in a shop, "we struggle to find things to keep the children busy," confessed one Old Order parent. Against this backdrop of parental concern about healthy alternatives to competitive sports and excessive partying, the transformation of older forms of outdoor recreation and the emergence of new ones take on significance. Many Amish trace the intensification of outdoor recreational activities and sports to the

1970s, precisely the time when many shop workers were looking for ways to maintain their connection to the land.

This chapter explores longstanding outdoor provisioning activities, such as hunting and fishing, as well as newer pastimes, such as birding and horseback riding, in an effort to understand the changing meanings of nature for Amish outdoor enthusiasts. We also ask how the Amish organize their outdoor hobbies and explore Amish attitudes toward and responses to "the regulatory state," which develops rules and guidelines for outdoor pursuits at various levels of government.

Outdoor Recreation and Provisioning

In urban and suburban settings, harvesting wild game and food through hunting, fishing, trapping, or gathering is widely seen as an anachronism because it is no longer necessary for survival. For the Amish, however, as for many rural English, these outdoor activities not only contribute food for the table but create special bonds between family members. Outdoor adventures also solidify attachments to particular landscapes, where memories are created and passed down in family lore. Yet Amish outdoor recreational activities have received little coverage in scholarly works, and the idea that leisure is inappropriate lingers in some Amish circles. "In our world, free time and recreation are seen as negative," commented one New Order man. Times are changing, however, and more and more Amish see outdoor recreational activities in a positive light as long as they are kept in proportion and remain "good, healthy fun."

Hunting

It is hard to overstate the significance of deer hunting in many Amish communities around the country. The opening day of deer season may have a holiday feel, as anticipation percolates through extended families and conversations about hunting dominate church, school, and workplace. In many settlements an entire industry has sprung up around deer hunting, including Amish businesses that specialize in outdoor equipment, taxidermy, custom meat processing, deer feeders and feed plots, items made from deer antlers, and more. For one Old Order man, an investment of a thousand dollars in a bowstring business has grown into a very profitable company with many hundreds of dealers across the globe.[3] By contrast, a Swartzentruber sawmill owner makes about twenty wooden, treehouselike deer blinds annually, including a $450 deluxe model with tinted windows and "room for a table and swivel chairs so they can play cards." Amish regularly attend regional sportsman shows, and many households subscribe to hunting magazines that offer expert advice on everything from scent removal to tree-stand placement.

Most Amish hunters do not skimp on technology, including firearms, in the pursuit of their quarry. While the Amish are well known for their refusal to bear arms against other people,[4] they have no qualms about owning and using guns for hunting.[5] A few Amish even own handguns for killing trapped animals or for self-defense, especially out West, where "it's standard to carry." Most households have a shotgun or a .22-caliber rifle for exterminating animals they consider pests, and hunting rifles are among Amish men's most treasured possessions.[6] And in spite of the occasional hunting accident, a wildlife official we spoke with felt that the Amish work more on their shooting skills than the English. "It's instilled in them at a young age to be a good shot because they don't want animals to suffer."

Nor do Amish hunters necessarily refrain from incorporating photography into their outdoor adventures. Many a successful hunt has been captured by the Amish on a disposable camera or by a non-Amish driver or friend. Though leaders in some church districts still frown on their use, Amish hunters often mount battery-powered cameras in woodlots they own or rent to track the movements of deer, especially big bucks, in the off-season. "Trail cameras—just about any Amish guy will use them," commented an Old Order hunter. The relative acceptability of trail cameras rests on the fact that they are trained on nature, not people, that the photos are not used for commercial purposes, and that the cameras are used on properties far from the public eye.

Like any hobbyists, Amish hunters vary, ranging from the casual hunter who saunters into his back woods a few evenings a year to those who leave no stone unturned in their quest for a trophy buck. In general, though, Amish hunters have a reputation for "going all out." One Amish hunter in his late twenties from Topeka, Indiana, who leases 110 acres of hunting land for twenty-five hundred dollars a year, goes out weekly in the months before hunting season to observe which trails and bedding areas deer use and then to put up tree stands accordingly. He also plants a plot of alfalfa and scatters Lucky Buck mineral supplement on the ground to attract deer. Most Amish hunters abandon their usual attire in order to wear camouflage, spray themselves down with a scent remover, and try to attract deer by rattling antlers or using commercially available "tube" and "can" calls that imitate the grunting of bucks or the bleating of does and fawns.[7]

Hunting in the United States is a predominantly masculine sport, with one national survey estimating that 91 percent of hunters are males.[8] The Amish are no

Although some Amish still hunt deer locally with minimal technology, others buy or rent land just for deer hunting and use trail cameras and feeding hoppers to monitor and attract deer. Photo by Doyle Yoder

exception to this trend. "When I was growing up, if you didn't hunt, it was almost as if you didn't belong," reflected one Old Order father of ten. Shooting one's first deer is akin to a rite of passage for many boys. Many fathers emphasized the special bonds forged with their sons while hunting deer. The competitive and masculine dimensions of hunting are magnified in western Amish settlements because of the available game. In fact the Amish community in Rexford, Montana, has become known as a kind of bachelor's paradise in part because of the hunting opportunities. "It's more important than their jobs," recounted one man who grew up there. Many hunters we spoke with described the sense of self-sufficiency that comes from a successful hunt. "There is a satisfaction with bringing home the meat," one Old Order man pointed out. Then he laughed, "I'd like to be praised for bringing home the meat, but my wife doesn't like venison!"

In non-Amish circles, female deer hunters are more likely to embrace hunting after they are married, as an activity important to their spouses.[9] In contrast, Amish women who hunt usually start hunting as children with their fathers, or with their boyfriends while dating, and then cut back sharply once they have children. "We as a family have great memories of the girls going along deer hunting," recounted a man whose daughter shot two deer in a period of two weeks.[10]

Even so, Amish women in the Midwest may be regarded as "kind of odd" if they continue to hunt after motherhood. In western Amish settlements, however, "most of the ladies still hunt, and they take just as much pride in it as the men," confided a mother of six who was nursing a new baby. Asked to compare hunting in Rexford with hunting back East, she replied, "Oh, it's way different. They don't know what we're talking about when we say we go looking for sheds."[11]

Like most English hunters, the Amish respect the deer and revel in the thrill of the chase. In recent years, as the technology surrounding crossbows and compound bows has dramatically improved, the Amish have taken up bow hunting in large numbers, in part because they say there's something extremely satisfying, even primordial, about killing a deer with a bow and arrow. Asked if he ever felt sad about killing a deer, one father replied, "So far I guess I haven't. That's just the way I was brought up. But if I get a deer, I always stop and think about how lucky we are to have things like this available." A mother and hunter in Montana reflected, "Well, you kind of feel bad that you're killing a beautiful animal, but elk can really make a mess out of our hay fields." Amish hunters do not subscribe to the notion of a balanced nature; they believe that if humans don't keep nature in balance, certain plants and animals would overrun others. "Yeah, if deer wouldn't be hunted, it would be a crime," said one man. "I mean we would probably have them right out here in the yard."

At the same time, the Amish are engaged in a lively debate among themselves about when deer hunting crosses the line in terms of the obsession with antlers and the time and money spent. One writer to the conservative Amish publication *Plain Interest* described a hypothetical hunter who spent all of late summer scouting and buying gear, after which he "looks like he might have stepped out of a *Cabelas* catalog."[12] He hires a driver on opening day, and though he harvests a buck, he arrives home exhausted, long after dark. The author's critique indirectly acknowledges debate within the Amish community about what counts as a wife's "reasonable support" for her husband's hobby. "Some wives have a problem with husbands being gone for weeks and some don't," noted one hunter, whose friend had "quit hunting because of his wife."

Though whitetail deer are their main quarry, Amish hunters pursue other game as well. "There's nothing like a nice spring morning with turkeys gobbling," said one man. Some specialize in particular kinds of hunting, such as using dogs to track rabbits or coyotes in the winter. Amish men living out West hunt elk, mule deer, and mountain lions, and they often enter annual lotteries seeking a permit to hunt more highly regulated big-game animals, such as bighorn sheep or antelope. One shop owner in La Jara, Colorado, displays in his front office a full body mount of a cougar he shot. Amish-authored hunting magazines and memoirs are replete with stories of small groups of men who travel to Wyoming or Idaho and go on a guided hunt. Guided trips to hunt black bears in Canada are not uncommon, though "there was a guy who missed communion for bear hunting, and something was said about that," an Old Order Ohio man confessed. Even so, hunting trips to exotic locations in pursuit of a trophy rack provide a welcome respite from the hectic pace of shop culture. Like hunting trips to favorite local sites, they are also highly valued for the social camaraderie they afford in an all-male environment.

Fishing, Trapping, and Foraging

Amish provisioning is by no means limited to hunting, as our early-morning visit in late February 2012 to Mrs. Yoder's Kitchen in Mt. Hope, Ohio, attested. For nearly fifteen consecutive years, the annual Fisherman's Breakfast has attracted many dozens of Amish fishermen to hear an update from state wildlife officials about the Lake Erie fishing scene. Coming after one of the worst years of walleye fishing in recent memory, the discussion that morning focused on the effects of the green algal bloom in Lake Erie and whether commercial fishing was cutting into sport fishing. Most of the members of the nearly all-Amish audience were boat owners; some hired drivers to pull their boats, others "parked" a boat on the Lake Erie shore or at an inland lake, and a few even ran their own charter-boat businesses. According to the Amish organizer of the Fisherman's Breakfast, "It's all about wanting to make fishing on Lake Erie better. In 2004 the fishing got tough and I asked my friends, what can we do to make sure Lake Erie stays healthy? It was like we were in the dark."

The growth in fishing among the Amish goes back to the 1970s and the rise in hourly jobs, which gave men Saturdays off. An Old Order man recalled that when he first rented a boat in 1975, apart from catching bass and blue gill in farm ponds, "very few Amish guys went fishing." Since that time, fishing has changed greatly.

One English outdoor enthusiast who often fishes with Amish friends said, "The first Amish guy I took fishing, I expected a cane pole and a few lures, but he had a tackle box that put me to shame. Another guy had a $40,000 boat with a $20,000 motor and GPS." For serious Amish fishermen in the Midwest, outings to Lake Michigan, Lake Ontario, or sizeable inland lakes are not uncommon, and several Amish-organized charity fishing tournaments are held annually. A Lake Erie walleye tournament in its twentieth year raised five thousand dollars for medical bills for a needy patient in 2014. It attracted 31 boats and 170 Amish participants who did drift fishing and weighed in one fish per angler to see which boat had the highest average.[13]

Amish fishermen are not lollygaggers. While "there are guys who will 'fun fish,'" as one person put it, catch and release is rare in Amish circles. "Usually, if I go fishing, I like to get a meal to eat," commented an Andy Weaver father of eight. "An Amish man doesn't spend his money where he doesn't have a return," observed one avid fisherman. "We don't go fishing to drink beer. We can do that at home," he joked. Amish fishermen pay careful attention to the seasons, the water temperature, spawning, and which fish are biting where. "I have plenty of weapons, that's what I call 'em," said one Amish man who makes his own lures. Because Amish fishermen have to pay a driver, it's less expensive to split the cost and more fun to go fishing with a group. One man's fishing trip involved a 4:00 a.m. departure, a three-hour drive to the lake, eleven hours of nonstop fishing, in which the entire group caught their limit of perch, and a return home by 1:00 a.m.

Trapping is another form of provisioning that attracts some Amish boys who enjoy the challenge of outwitting a wild creature and the chance to make some pocket money by selling furs. Though the market for furs is relatively weak, it's not uncommon to see coon, muskrat, and fox pelts tacked up to dry in Amish barns. In September 2011 we attended a day-long trapping seminar in Baltic, Ohio, hosted by an Old Order Amish man nicknamed "Fur Al" that helped nearly two dozen nine- to sixteen-year-old Amish boys complete their state-mandated trapper-education course.[14] The Ohio Department of Natural Resources official who conducted this workshop confided, "The Amish grade you. If they like you, they'll ask you back." One reason Amish trappers like this presenter is that he shares their critical view of animal-rights activists. According to one Amish man, in 1977 Ohio Amish turned out in higher than usual numbers to vote against a constitutional amendment banning leg-hold traps. Attitudes are slowly changing, however. "I used to trap," said an Amish bishop. "Now I'm becoming soft-hearted. The challenge was to get that sly animal to step in a one-inch piece of metal in a ten-acre field. It's just not worth it." Most Amish who trap, however, see themselves as doing a service for the community by preventing an overabundance of pests.

Foraging for nuts, berries, and mushrooms is another popular outdoor activity for many Amish families, and favorite locations are closely guarded secrets. "Where'd you get your morels? That's like asking, 'How much money do you have in the bank today?'" laughed a New Order minister. An Amish man from Big Valley, Pennsylvania, remembered how he and his dad had once found morels that were so plentiful that they filled not only their bags but also their hats and their denim jackets because they just couldn't bear to leave any behind. In the western Amish settlements, families keep close tabs on tracts of land that have been burned by forest fires as a potential gold mine for morels the following year. "We took two vanloads," said one New Order bishop in Montana, who showed us leftovers in his refrigerator from four gallons he had harvested. "Back home you count every morel you find," said one man, "but out here you can fill shopping bags full of them."

Amish Outdoor Provisioning and the Regulatory Context

Amish hunting and harvesting takes place within a complex web of federal, state, and local regulations aimed at balancing the recreational use of natural resources with the long-term health of plant and animal species. According to an Ohio Division of Natural Resources wildlife official, it is not uncommon for English outdoor enthusiasts to badmouth the Amish, accusing them of spotlighting deer and stretching the regulations. "'The Amish moved in, and now there's no deer.' We hear that a lot." One practice that may contribute to this perception, he said, is that of vanloads of Amish from the Geauga settlement, east of Cleveland, traveling to southern Ohio to hunt deer: these hunters position "standers," who walk the woods and push the deer to "shooters" stationed along a road or a field.[15] Although hunting in groups and from roads is practiced by only a small minority of Amish and English hunters, it is the Amish who often attract attention because of their appearance.

Like their English counterparts, the Amish occasionally harvest wild game without the proper permits. Yet the wildlife officials we spoke with were generally complimentary about Amish compliance with regulations. "I have very few complaints on the Amish," noted one game warden. "Most are respectful of the law, though we've got the whole spectrum. Some Amish kids will give us the finger and run the other way." An experienced English hunter concurred, "I don't know if there's any basis for [criticism of Amish hunters]. I think it's a prejudice against them." Several Amish outdoorsmen echoed this view that the Amish are held to a higher standard because even when they are wearing camo, their distinctive personal appearance often makes them identifiable in the outdoors.

Ecological Amish Meets Ecological Indian

The most interesting example we encountered of the cultural tug of war over regulations on outdoor provisioning comes from a New Order Amish settlement in St. Ignatius, Montana. The approximately thirty Amish families who live there are surrounded by land owned by the Confederated Salish and Kootenai Tribes of the Flathead Reservation. Created by the Treaty of Hellgate in 1855, the reservation encompassed 1.2 million acres in western Montana for enrolled members of the Bitterroot Salish, Pend d'Oreilles, and Kootenai tribes, whose ancestors were marched to the site from their homeland further east. In the early 1900s the US government declared portions of reservation land to be "surplus" and opened them to homesteading by white settlers. The result of this "checkerboarding policy" is that about one-third of the acreage in the Flathead Reservation is owned by people who are not tribal members.[16] The tribal council is trying to rectify this situation by buying back parcels of land as they come on the market, using proceeds from a dam, a casino, and other income-generating ventures.

After Amish families moved into St. Ignatius, in the Mission Valley, in 1997 and grew to more than forty households over the next fifteen years, conflict erupted between the two groups over natural-resource use. An employee in the Tribal Lands Department observed that "the regulation and use of forest land is where we bump up against [the Amish]. They've taken the liberty to take what they want. They use reservation forest as their backyard. They help themselves to our timber, fisheries, berries. They believe they can skate around all of it. They say they will [abide by the regulations], but action doesn't follow." Another tribal member concurred: "Some [of the Amish] are very respectful, but some think, 'It's God's country and we have every right to it.'" Of particular concern for tribal members has been the perception that the Amish take large groups up to mountain lakes and "fish them out" and that they forage without the required permits.[17] A longtime employee of the tribal Natural Resources Department told us that since the arrival of the Amish, the tribal office had been forced to reduce the daily limit on trout, require that all trout caught be eaten on site (no packing fish out), limit party size and length of stay in the backcountry, and ban the collection of huckleberries, morel mushrooms, or antler sheds on reservation land. He concluded, "The Amish stand out for the simple reason that we've had to adapt our regulations in response to their use."

In 2011, tensions reached the point that the tribal Culture Committee and the Natural Resources Department asked Amish elders to attend a meeting to discuss the "alleged overharvesting of berries and fish by the Amish within the exterior boundaries of the reservation."[18] At this meeting Tony Incashola, chair of the tribal Culture Committee, explained the history of injustices experienced by the tribe and their philosophy of resource management: "Our ancestors protected the land and the resources for us. It is our obligation to do the same. . . . All living things are created equal and need to be treated with respect." Another elder commented, "American Indians have long understood the finite nature of nature—take what is needed and leave the rest for the needs of others," and the game warden reiterated, "You need permits to be in the woods."

The Amish elders who attended the meeting, however, responded that they were being wrongfully accused: "We want to live among you in peacefulness. We also know we are being blamed for things we didn't do," commented one leader. "This has been a bit of a culture clash." The Amish we spoke with felt that they were generally compliant with the law and that they were being singled out by tribal members who couldn't tell the difference between the Amish, the German Baptists, the Mormons, and other religious groups living in the valley. When tribal elders requested another meeting with the Amish several years later to air their differences yet again, the Amish leaders flatly refused. "We told them it was discrimination to call a meeting just for our group." Since then, the Amish and tribal members have continued to engage in limited commercial transactions, such as the buying and selling of timber or the processing of wild game, but an undercurrent of mutual suspicion lingers around recreational land use.

For their part, the Amish we talked with often recognized the historical injustices suffered by native peoples. "It's a tough situation. From way back they were mistreated and made to be dependent on the government," commented one man. Yet many of the stereotypes about native peoples that circulate in the wider society find fertile ground among the Amish in St. Ignatius. "Their goal is they still want all the land and all the whites off," said one man. Other Amish told us that tribal members are "pretty much spoiled," use "lots of drugs and alcohol," and "like stealing from whites." Many Amish in St. Ignatius view the native philosophy of land management as misguided in the sense that it elevates the status of nonhuman life to an unreasonable level. "If you shoot a grizzly bear in self-defense, they're sad about the grizzly," said one Amish man in disbelief. His comment makes clear the contrast between his anthropocentric approach to nature and a tribal culture that has long posited a sacred spirit in certain animals, including grizzly bears.

The perception that tribal members enforce restrictions and yet hardly use reservation land themselves rankles many Amish. One man reflected, "The biggest difference is the work ethic. Our group is all over the mountains, but their group is not. It's too energetic [for them]." Another elaborated, "We live closer to the land than the young tribal members do. The Amish have bigger gardens and they hunt more. Tribal members will hunt from the road." An Amish leader described tribal members this way: "The old people still value the culture, and they like to preserve the mountains. But the young people don't get outdoors as much. What we read in the paper about their connection to the land, we don't see very much. The elders see us out there all the time and wish their young people were. That may be behind their animosity toward us."[19] The tendency of the Amish to hunt, fish, and gather "in bunches" especially irks tribal members. Because they have large families and frequent visitors from back East and must hire and pay a driver, the Amish see traveling to outdoor recreational sites by the vanload as economical. But since their distinct dress makes them stand out, this practice only confirms the tribal perspective that they are a clannish community. "They're good stewards of their own land, but they're not team players," commented the head of the Tribal Lands Department. " 'Managing well' stops right at their boundary, and when it's somebody else's land, they're not as mindful."

We spoke with a tribal game warden to ascertain the extent of Amish infractions of tribal rules. "When they first arrived, we had a lot of infractions," he said. "For a while they kind of fought it. They didn't believe in buying permits." Several Amish we spoke with concurred: "They had a couple of meetings with us. They probably had reason to do that because we weren't aware of all the regulations," noted one. Over time, however, the situation has improved.

Nevertheless, several long-term Amish residents of St. Ignatius saw a connection between friction with the tribe over land use and the fact that some families have left the settlement over the past five years. "That's why some men move off," confided one Amish community member. "They don't like that part of it." Ironically, then, two groups whose public images are closely tied with stewardship of the land have not been able to see eye to eye on the management and harvesting of natural resources in a shared landscape.

New Outdoor Hobbies: Birding

On the evening of June 7, 2014, Henry Troyer Jr. was mowing hay on his eighty-acre farm in Coshocton County, Ohio, when the blade plugged up and he stopped the horses. At that moment, he heard an unfamiliar bird call from the uncut hay in front of him. Over the years, his extended family had kept a list, totaling 199, of bird species seen from their farm, and he immediately sensed that number 200 could be the bird in front of him. He rushed home and told his sons and his nephews, who brought over an electronic bird call. Just before dusk, they finally saw it: a rare bird known as a black rail. Troyer could hardly sleep that night, and "the following week was pretty crazy." Nearly four hundred birders, Amish and English alike, came to his farm from all over the Midwest to try to see the rail. An even bigger surprise came six weeks later, though, when he discovered one adult and two fuzzy black chicks, the first confirmed nesting of a black rail in Ohio. "Unbelievable!" he reflected. "To me, this was even more exciting than finding the rail initially!"[20] The Troyer family's experience of finding and reporting a rare bird and then hosting hundreds of birders who came to see it has been replicated by dozens of other Amish families who have sighted rare birds.

In one sense, close observation of birds has long been part and parcel of Amish farm and rural life. The Amish know that barn swallows regularly nest under the eaves of outbuildings, that killdeer frequent plowed fields, and that bluebirds can be coaxed with nest boxes placed along the edges of fields. In addition, the Amish have long held a special affinity for purple martins, a large swallow that nests in colonies close to houses and whose appetite for insects and reassuring twitters make for good neighbors. Many with suitable habitat around their homes have installed martin houses and take joy in the annual spring ritual of waiting for the martins to return.

In tandem with its growing popularity in the wider population, however, Amish birding has morphed into a significantly different form than bird observation on the farm. "When I started birding in 1988, if you went out and looked up in the sky with binoculars, people wondered what you were doing," remembered one man. Most Amish birders trace the origins of their hobby to influential Mennonite and Amish mentors in the Holmes County, Ohio, settlement, but it has now spread, though somewhat unevenly, to Amish settlements across the country. Like hunting and fishing, birding attracts primarily boys and men, who use the latest Swarovski or Leica optics in "keen but friendly competition." Amish birders are very attuned to habitats, songs, plumages, and behaviors, and they bird at an intense pace. One avid Old Order birder put it this way: "I'm not saying that non-Amish people can't be good birders, it's just that the majority of the city people will go birding in a more low-key way. We go birding like how we sincerely do our work, you know. We get every bird that we can." Like English birders, the Amish keep lists of how many species they've seen in their lifetime in North America, in a given state or

Amish birders travel far and wide in search of species for their year and life lists, as these women on Sanibel Island, Florida, attest. Photo by Leticia Swam

county, and on their own property. "I thought about starting a list of birds I heard in church, but decided people might think I wasn't paying attention," quipped one birder. "But I had six warblers one morning."

In northeastern Ohio, Amish birders turn up rarities at an unusually high rate given that the areas where they live lack coastal or mountainous habitats. One reason is that Amish lands tend to be relatively undeveloped and managed in bird-friendly ways, in contrast to the glaciated western and northern parts of Ohio, where agribusiness has created monocultures or urbanization has led to cement deserts. But according to a well-respected non-Amish naturalist, "The real reason that so many rare birds turn up so consistently in this area is the people. Some 40,000 Amish reside in this region, and their ranks include some of the best birders in the state. Also, a high percentage of the Amish community is far more bird-literate than the average population, so there are lots of knowledgeable eyes and ears out on the ground."[21] Amish birders' careful observation of when and where birds are likely to show up has even gained national attention: they were featured in a 2017 article in *Bird Watcher's Digest*.[22] Their finds have contributed to the ornithological knowledge base about bird distribution and behavior and have led to a high degree of cross-communication with the English birding community.

At the same time, Amish birders have demonstrated a remarkably creative capacity to mold this hobby to fit their own lifestyle and sensibilities. Using answering machines in their phones shanties, Amish individuals operate very effective rare-bird alerts in multiple states at a time when such alerts have largely migrated to listservs and social media for the non-Amish. Their birding competitions often focus on year lists or even a monthly challenge, such as a "big November," which require careful attention to seasonal changes and migration patterns. A "bicycle big year" has become highly competitive among young Amish men, who vie to record the most birds seen in a year in the area around their home that can be reached by bicycle.[23] Amish editors publish two national birding magazines, *Feathers and Friends* and *Journey of Wings*, which rely largely on letters from Amish birders sharing their sightings and stories from the field. Finally, the Amish are active participants in local Christmas bird counts sponsored by the Audubon Society, which are used to track bird numbers nationally. In all these ways, they shape birding so that it resonates with their values and their use of technology, molding it into an attractive hobby for children, youth, and adults.

Occasionally, aspects of the English birding world collide with the religious beliefs of Amish birders. One Amish birder we spoke with expressed surprise after being told by a well-known naturalist that most English birders probably are not Christians. A highly respected Amish birder who declined when he was asked to serve on his state's rare-bird records committee explained: "I would have been qualified, probably, and I was interested, but I wouldn't want to be on a committee with people that don't have the same values and faith that I do." One of those values is clearly spelled out in the faith-based mission statement for an Amish birding magazine: "The *Journey of Wings* staff and columnists seek to teach the real origin of the earth, giving glory and honor to God. Who is the Creator of the whole universe and everything that is in it—including the birds we love." In their efforts to keep a degree of separation from the world, some Amish birders thus selectively choose not to align themselves officially with groups who do not share a biblical world-view.

Another interesting example of where Amish birders draw a different line than their English counterparts is in their reaction to the annual update to the list of all known birds in North America and in the world, referred to as the bird taxonomy. The changes have practical consequences for the number of birds on one's life list, because each year a handful of bird species may be split into two, or conversely, what were previously two species of birds may be lumped into one. From a scientific perspective, these changes have come as DNA evidence has helped biologists to better understand the evolutionary relationships between bird species. But while Amish birders do incorporate the updates into their lists, they largely dismiss the logic behind the changes as reflecting the whims of scientists. "I think it's kind of ridiculous," said one Old Order man. *Journey of Wings* even distributes a checklist to its subscribers that does not follow taxonomic order but instead divides the avian world into the "commonsense" categories of water birds and land birds.

In spite of these differences, Amish birders forge surprisingly strong friendships with Amish from other affiliations and with English birders, perhaps because the competition is more subdued than in hunting or fishing. "If us three are birders and I see an Ivory-billed [woodpecker], I'll let you know, right? But if we see a big buck, we're gonna protect that with tooth and nail," commented one man. Amish birders note that hunters view them as "soft," which often leads to good-natured ribbing, but the birders also see their hobby as somewhat cleaner and purer than hunting. According to one businessman, "Birdwatching is not seen as a sport. Even though men spend quite a bit of money doing it. Because you don't harvest anything. You're an observer, but not a hunter-gatherer. Rarely is birdwatching, or the excesses of it, censored in our community. It's good, clean fun, and it sure beats drugs and alcohol and fornication." Nevertheless, like Amish men who hunt, birders do worry about the cost and about the time spent away from their families. "If they knew what I was doing, most of the members of my church would think I'm absolutely crazy to go chase a bird all day," confessed one Old Order man.

Recreational Horse Riding and Training

On a gorgeous fall morning in late September 2016 the Eighteenth Annual Benefit Trail Ride kicked off near Mount Eaton, Ohio, to raise money for medical expenses for a wheelchair patient. Just over 250 riders, mostly Amish, traversed a nine-mile trail across contiguous properties for three hours before arriving back at the host's farm, where a makeshift corral, benches, and food tent had been set up. The rider who raised the most money from sponsors, which was more than thirteen thousand dollars, received a saddle as a prize, while the two runners-up chose between a mountain bike and a deer stand. After everyone had eaten lunch, prepared by volunteers on site, the emcee led the riders through several hours of competitions: electronically timed barrel races, pole bending, and water relays, as well as jumps. In the Kid's Stick Horse Race, toddlers and young children scampered twenty yards on wooden stick horses, receiving much laughter, applause, and candy for their efforts.

Concerns about modesty prevent some Amish church districts from supporting organized horseback riding by girls, but Old Order parents in some settlements see benefit trail rides as a healthy outdoor recreational outlet. Photo by Doyle Yoder

Trail rides are a relatively recent phenomenon, dating back only to the early 1990s, when an Ohio Amish man started the Salt Fork Trail Ride as a healthy recreational alternative for youths fifteen and over. At that time "there was not a whole lot of people riding recreationally," said one adult who accompanied fifty-five young people on the overnight outing. Since then, trail rides have found a niche among a segment of the Old Order in Ohio and Indiana as a wholesome recreational activity consistent with the historical importance of horses in Amish life. The program of a benefit trail ride varies depending on the organizers and the cause. The Bristol Ride in northern Indiana, for example, is usually a two-day affair, with a short Friday night ride followed by dinner, a cornhole or volleyball tournament, and a guest entertainer, with an option to camp out before the longer ride on Saturday. "When I started, there were maybe fifty people I knew that were horseback riding," said the organizer of the Bristol Ride. "Now there's probably about 150, 200 of them. We've got a lot of youth."

Many of those youths are girls. Horseback riding stands out as an outdoor recreational activity that some parents encourage their daughters to pursue. "I know for sure it's very popular," said one Old Order mother in Holmes County, Ohio. "In some areas, all the girls have riding horses." Many young girls start with a pony and cart, and then move up to a riding horse. One horse trainer thought that girls were better riders than boys because they accepted advice more willingly. The concern about modesty, however, has prevented riding culture from being adopted by some Amish. Because girls typically ride in jeans or sweatpants worn under their dresses and might mix with boys on a trail ride, "horseback riding is not encouraged in our churches," said a New Order minister in Ohio. One Ohio horse trainer noted that "Pennsylvania is really slow as far as coming along with riding," while ultra-conservative Amish groups rarely get involved in recreational riding. Beyond the issue of modesty, leaders in the more conservative churches object to the ways the "higher" Amish use ponies and riding horses as a status symbol and, in their view, justify recreational riding by turning them into benefit rides.

Amish horseback riders exhibit a strong preference for Western as opposed to English riding style. This means bigger, heavier saddles with a pommel, designed for comfort and utility, as well as split reins, which are held with one hand, not two. Though few Amish in the Midwest use horses to round up cows, Western riding style makes sense for Amish living in Montana, where beef, not dairy, cattle are an economic mainstay. The various competitions at trail rides and at events that celebrate horse culture, such as Horse Progress Days, also follow the Western riding style, as does the newest dimension of Amish horseback riding culture, drill teams. We first encountered one of these teams, the Saddle-Up Cowgirls, at a benefit auction and trail ride to help with medical bills for a close friend and were mesmerized by the choreographed performance of the thirteen- to fifteen-year-old girls, four dressed in blue and four in pink, with matching leggings on their horses.[24] These young riders and those who support them are redefining what is considered an acceptable public performance and outdoor recreational pursuit for women.

The steady growth of riding culture has both reflected and promoted a broader shift in the Amish relationship with horses, especially the embrace of "natural horsemanship."[25] "To train a horse is not called breaking a horse anymore," said one horse trainer, who added that in the past, "if the horse wasn't traffic safe, Grandpa would have tied him by a busy highway." The new approach, which many Amish attribute to the influence of English horse trainers such as Buck Brannaman and Monty Roberts, involves building a relationship with your horse and gaining its trust rather than forcing it into submission. "We use the imprint training method," said one Old Order man, referring to veterinarian Robert Miller's horse-centric approach, which emphasizes rubbing and touching the foal immediately after birth. "We would look at it as the horse is happiest if well-fed. I always had a rule, the horse eats before we do." The Amish are not immune, then, to the changing cultural milieu, which, as the psychologist Richard Serpell argues, has seen "the development of a more egalitarian approach to animals and the natural world."[26] Still, old habits die hard. An experienced horse trainer observed that the move to a less coercive approach still "has a far ways to go." A non-Amish equine therapy specialist who works with Amish clients concurred, saying that she sees a mix of old and new techniques and that some Amish still insist on controlling the horse physically to keep it still.

Their growing recreational use has also increased sentimental attachments to horses. After all, it's easier to distance oneself and one's emotions from a wild animal that is only fleetingly observed or even a domesticated animal that provides food, labor, or transportation.[27] "You get attached to them, and it's really hard," said one horse trainer. When a horse nears the end of its normal life span, however, most Amish have few qualms about taking it to a horse auction for disposal. "We really feel when a horse is old it should be put down in a humane way," reflected one horse trainer. "[PETA] really looks down upon us for that, but there's not a hereafter for a horse, and it's more cruel [to let him live out his life] than putting him down." Viewed broadly, then, the recreational use of horses has allowed some Amish families to keep horses as a central part of their lives in spite of the decline of farming and has increased the acceptability of new kinds of sentimentalities and public performances. Rising medical costs, which can be defrayed through benefit trail rides, and concerns about clean activities for the youth are catalysts for these activities and provide a "kind of justification," said one Amish man, who felt that recreational riding was "meeting a need" for leisure that reflected the changing occupational structures of Amish life.

Preserving Cultural Autonomy in the Outdoors

Few ethnic or religious groups in contemporary North America remain as closely tied to nature for provisioning and for recreation as the Amish. Though the rise of shop culture has changed their scope and form, the pursuit of wild game and the enjoyment of outdoor activities is still central to Amish life. At the same time, the Amish tend to hunt, fish, trap, gather, bird, and ride horses on their own terms. They voraciously gather outside information but screen and channel it for their purposes. They fashion competitions and charity events that are suited to their values and interests. They create small businesses that support and nurture their outdoor hobbies. In none of these pursuits are non-Amish unwelcome; to the contrary, the

Amish rely heavily on English drivers, consult with English outdoor experts, and accompany English friends who share their interests. Still, Amish provisioning and outdoor recreation reflect their deliberate marginality from the outside world, and this degree of autonomy sometimes leads the Amish to act as if they were living in a self-contained world. "We tend to be clannish," admitted one Amish opinion leader. Another man reflected, "Now that I come to think of it, Amish people can be pretty defensive on what is theirs. Maybe even too much so in nature because we kind of do our thing." The impulse toward insularity that has been crucial to the survival and growth of the Amish population as a whole, therefore, leads Amish outdoor enthusiasts to create a parallel universe of practice that overlaps and intersects with English outdoor recreational practices in important respects but also preserves a substantial degree of cultural autonomy.

CHAPTER NINE

Observing and Writing Nature

Amish as Travelers and Authors

There once was a man called Leroy
Took his wife, two girls and his boy
And Dave drove the van
He's a jolly good man
On a road trip they all would enjoy.

Leroy said, "Dave, here's the thing.
Let's travel south in the spring.
Wouldn't it be funner
To see a Texas roadrunner
Than any bird Ohio can bring?"

So begins a set of limericks composed by an Old Order Amish woman at the start of her family's three-week vacation to Texas and Arizona in the spring of 2012. Most outsiders think of the Amish as relatively immobile, restricted in their encounters with the wider world. Yet the reality is that the Amish are traveling as never before, often to sites of natural interest.[1] While the phrase "Amish tourism" usually conjures up images of nostalgic non-Amish tourists descending on "Amish Country," it could just as well refer to the thousands of Amish who annually visit such natural wonders as the Great Smoky Mountains and Yellowstone or who seek out nature while visiting relatives in other settlements. This burgeoning Amish travel industry, a product of growing levels of affluence and leisure, has the potential to foster awareness of landscapes and environmental issues that stretch far beyond Amish doorsteps.

Thirty years ago, Amish travels to other parts of North America would have been shared with friends, relatives, and church members primarily by word of mouth or recorded in a personal diary or a circle letter.[2] In addition to these personal conversations, however, Amish today are writing about nature in an increasing number of newsletters, periodicals, and books. And a small but growing number of artists are representing people, animals, and landscapes in new forms of artwork that push the boundaries of their respective Ordnungs. This outpouring of "Amish expression" is geared largely, though not entirely, to an Amish audience and is driven by Old and New Order men and women in the larger settlements. More conservative Amish still resist some of these written or artistic works as too worldly and individualistic. But according to one Old Order writer, "Those of us who have an appreciation for the fine arts are able to shrug off those restraints."

Amish choices about travel destinations, writing topics and styles, and artistic pursuits are deliberately selective, however. In this chapter, we explore how Amish travelers, authors, and artists filter their experiences of nature through their cultural and religious lenses. We begin by considering the meaning and role of nature in Amish travel and then analyze how Amish authors and artists write about and depict nature on the basis of their daily lives and experiences.

The Meaning and Role of Nature in Amish Travel

For some Amish families, travel outside their immediate locality is rare owing to occupation, family size, or financial constraints. Farmers are usually tied down throughout the year, or at least until their children are old enough to take care of the animals. "We almost never go away," said one Old Order mother whose family milked thirty-four cows on twenty-four acres of land. The life stage of a family also plays a role, with older couples more likely to have the time and money to travel. Some families simply cannot afford it—one mother noted that there was "no money left over" after all the bills were paid—while those in the most conservative groups still feel that travel for leisure is "an unnecessary way to spend money."

The rising overall affluence of Amish communities, however, has created opportunities for those who want to get away from home, and some members of more liberal affiliations have broken openly with the cultural restraints around vacations. Many Amish now see a need for occasional periods of relaxation in which their ordinary routines are temporarily suspended. For Amish travelers, as for many American vacationers, it is not uncommon to seek out some degree of pampering, becoming "queens or kings for a day."[3] Asked what surprised him the most about Amish tourists, one tour-company worker laughed, "They're very demanding. They want it the way they want it." A driver of a busload of Amish women on a mother-daughter trip recalls being told, "We want to get home late enough that we don't have to cook supper, or do the dishes. So don't drive too fast!"

One important motive in contemporary tourism is the search for destinations that are perceived as purer, simpler, and more authentic. The tourism scholar Dean MacCannell ties this "search for authenticity" to the rise of a leisure class who seek to fill a void in their own lives and overcome "the shallowness of modernity."[4] While this may be an apt description of some English tourists visiting Amish settlements,[5] it does not adequately capture Amish motivations for travel. Their community provides the Amish with deep and authentic relationships, so they rarely think of tourism as a way to replace their current lives with something different that gives them meaning. Rather, the challenge for Amish tourists almost always involves balancing their newfound purchasing power and leisure time with being true to the spirit of their church district's religious and cultural principles.

Visiting sites of natural interest holds a strong appeal for Amish travelers precisely because it helps them reconcile these two competing forces. The desire to see the wonders of God's creation is beyond reproach from the church. Apart from entrance fees to the parks, the price is also right: one can see a beautiful waterfall or take a scenic hike for free. In addition, visiting a natural area minimizes unfamiliar and ambiguous situations. Traveling outside one's comfort zone creates a certain degree of anxiety for just about anyone, but this is especially true for the Amish. Their choice to wear distinctive clothing marks them as a highly "visible" religious minority any time they set foot outside their vehicles.[6] Though traveling to natural sites does not guarantee that they won't be objects of curiosity, such visits are generally consistent with outsider expectations of Amish behavior.

Since the Amish do not own or drive motorized vehicles and most will not fly, their chosen modes of travel are also distinctive. Horse-and-buggy travel is practical only for short trips, but it is inexpensive and allows the occupants to be attuned to the rhythms of nature. Train or bus travel is also an option for all Amish, but these modes are used especially by conservative groups, whose travel is mostly limited to visiting members of their extended families. For vacation travel, however, most Amish join a chartered bus tour or hire a driver and vehicle. Both include the advantage of having non-Amish drivers, who are helpful in negotiating the logistics of travel. Several tour companies that specialize in charter bus trips to destinations across North America cater to an Amish clientele. In the case of vacations with a hired driver and vehicle, Amish families or groups of friends usually choose the destinations and types of activities, as well as their preferred lodging and meals, in consultation with the driver, who then makes the arrangements. These different modes of travel have important consequences for how one interacts with the outdoors on a trip.

Tourism of any sort is about identifying places as "sights" and visually consuming them, but it also involves moving through those places with other senses. Yet Amish travelers vary significantly in the time they spend "in nature" and in how they engage the landscape. Amish interactions with the natural world fall into three basic types: experiencing nature as backdrop, nature appreciation, and nature engagement. Though a given trip or a given traveler may interact with nature in more than one of these ways, travel experiences along this continuum are more or less likely to lead to in-depth engagements with the natural world.

Nature as Backdrop

Some Amish describe themselves as having only a passing interest in the natural world; for them, outdoor activities on any given trip may be peripheral at best. In addition to visiting relatives, many Amish travel for health care, for work, or on service trips. An Old Order family's attempts to find a cure for their daughter's unexplained illness led the mother and daughter to spend six weeks in an ocean-front apartment in Myrtle Beach, South Carolina, where they could see dolphins from the balcony. A furniture-store owner and his wife went to San Francisco and back by train; they enjoyed seeing the Rocky Mountains and other sights along the way, but the trip was paid for by a client, and its main purpose was business. A small number of New Order Amish have even flown to Central and South American countries for mission and outreach work, encountering landscapes very different from those back home. In such instances, encounters with nature and experiences that broaden their horizons are largely an unexpected by-product of travel, though they may still make a strong impression.

Warm temperatures and sandy beaches attract the many thousands of Amish snowbirds who each year flock to a small suburb of Sarasota, Florida, known as the Pinecraft settlement, to enjoy sun and fun. Pinecraft owes its origin to a visit to the area by a small group of Ohio Amish in the 1920s. Enamored with the fishing and hunting opportunities, they later returned to start a small settlement based on celery farming.[7] Though a few Amish still use Pinecraft as a launching point for deep-sea-fishing trips or birding excursions, most visitors are successful entrepreneurs and retirees from the liberal affiliations who see their stay as an escape from the cold weather and the hectic pace of life back home. The rules of their respective church districts are relaxed, and the most popular activities are visiting, bocce and shuffleboard, Scrabble and dominoes, and short trips to the beach at Siesta Key.[8]

Some Amish travelers prefer to visit the human-made buildings and infrastructure that mark a landscape. One family went to Maryland to visit an Amish friend who supplied leather for their business and decided to stay longer to visit the Washington Monument and the Supreme Court. While the Amish often visit zoos or aquariums, they are quite divided about visits to science and natural-history museums. Conservative Amish usually balk at museum visits of any kind because

National parks are prime destinations for Amish travelers who have the time and resources, as illustrated by these two couples at Glacier National Park. Photo by Luiz C. Ribeiro

portrayals of historical figures run counter to the biblical admonition against graven images. The way objects are displayed in museums also seems too close to idol worship for them. For higher groups, who usually don't take these biblical verses so literally, a major concern is that exhibits may mention or allude to long periods of earth's history. A popular destination since 2007 for some New and Old Order Amish has been the Creation Museum in Petersburg, Kentucky, which boasts visually appealing re-creations of the events in Genesis and a narrative that frames modern science through a creationist lens.[9] For trips with nature as a backdrop, contact with environments that are largely unmodified by humans is not a major drawing point.

Nature Appreciation

In contrast, some Amish travelers seek out beautiful scenery or famous natural sites but prefer not to spend too much time outdoors or give up any creature comforts. Nature appreciators sometimes combine travel for other reasons with stops at sites of natural interest. A father from the Nebraska Amish group in Big Valley, Pennsylvania, told us, "We put our feet in the Pacific Ocean eleven years ago. I saved my tobacco money and took my family, and they all enjoyed it." Over the course of a three-week trip that covered nine thousand miles, they stopped at national parks like Glacier and Yellowstone, but the real highlight was their visit to Colorado to see "where grandpa was born." Another extended family rented eight rooms at a motel in the Great Smoky Mountains between Christmas and New Year's Day. They enjoyed the scenery and took short walks to several waterfalls but also visited Gatlinburg and rode on the well-known Skylift and Mountain Roller Coaster. They appreciated the sites of natural beauty even though their engagement with the outdoors was largely structured by the tourist industry.

Organized tours to famous natural destinations strongly appeal to nature appreciators. Specializing in charter bus trips for a mostly Amish clientele, Green Country Tours, based in Mt. Hope, Ohio, includes lodging, meals, and admission to parks or events, as well as a guide who speaks Pennsylvania Dutch. "That's what's nice about Green Country Tours. Everything's taken care of," commented one bishop. Green Country offers tours to the Canadian Rockies, Niagara Falls, the Grand Canyon, and other spectacular natural sites across North America,[10] but the social aspects of traveling together are as important as learning about or experiencing the destination itself. "I think they enjoy each other's company more than anything else," said one Amish man. Participants are eager to find those with family connections among their fellow travelers, and the tour guide usually begins the trip by passing around a microphone for self-introductions. Group games are common, such as a balloon race from the front to the back of the bus. "One of the things that surprised me was how much they enjoy having fun. Like they are out to have a good time!" reflected a bus driver. Participants also bring back souvenirs, including "a lot of calendars with scenery pictures, coffee mugs or cups, and bottle openers," and they usually request a collection of photos of the sites visited, taken by the tour guide and uploaded onto a compact disc.

Even though organized tours offered by Green Country or a pilgrimage to Pinecraft allow for limited contact with natural sites, participants may still be moved by their encounters with the outdoors. A young woman who traveled with three friends to Mackinac Island, Michigan, wrote in her diary: "Then went back out to the lake. By now the moon was out and it was so beautiful! We just stood there and took it all in for a while—the peaceful sound of water moving in, ever-moving, never-silent, yet so tranquil. What a sight to behold with the moon on the water."[11] Nature appreciators may also be astute observers of changes in the climate and the landscape when they travel. An Old Order man who took a train trip out West reported that he and his wife were the only people looking out the window at the landscape; all the others were looking at their little screens.

Nature Engagement

A third subset of nature-related travel among the Amish involves both sustained interaction with the outdoors and an avid curiosity about how organisms fit into their respective ecosystems. Amish travelers who engage with natural settings in this deeper way are predominantly male and do not mind "roughing it." One Andy Weaver couple we interviewed had built a homemade box for the back of their buggy, which they used to take their family on summer camping trips "with children and gear sticking out all which ways." The father recollected, "We hardly used any tents. We trained 'em to sleep under the stars, and we'd get up all dewy and wet and stir up the campfire." An Old Order man recalled that his fondest memory from a trip to Glacier National Park was hiking up one of the tallest peaks with his boys. Three New Order young men on a trip to the Blue Ridge Mountains in Virginia got up early and literally ran the last mile to the top of a nearby peak, where they sat for an hour as the morning mist dissipated, observing birds and trees and speculating on the origin of the mountains in relation to the biblical flood.

Amish who want a deeper engagement with nature often seek out natural areas that are "less spoiled." In particular, those who live in western settlements are "way more interactive with the outdoors," as one man put it, and they often go on camping trips in the summer. Many Amish families who live in St. Ignatius, Montana, take a train of pack horses into the 1.5 million-acre Bob Marshall Wilderness, where they set up camp for a week or two of hiking and fishing. "I love to get up in the high country," said a New Order man in St. Ignatius. "The mountain wildflowers are just amazing. You feel a notch closer to God in a sense." Midwestern Amish who desire natural areas less impacted by humans have to travel further. An Ohio Amish man who took his family on an extended camping trip to the Upper Peninsula of Michigan noted, "If you want to get to wilderness, the closest place is to drive ten hours due north; otherwise, it's two days out west." This man realized that his version of a vacation was "not typical Amish," but for those who want sustained engagement with the outdoors, organized tours are too constraining, the bus rides filled with "silly humor," and the stops too commercialized.

Perhaps the most intense version of travel for nature engagement is travel in the single-minded pursuit of a quarry. The large majority of such trips are taken by men who are passionate about hunting, fishing, or birding. Often a group of age mates plan well in advance and save up for such a trip, which requires both mental focus and physical stamina. For example, a vanload of six Andy Weaver Amish from Apple Creek, Ohio, along with two drivers, took a ten-day trip to southeast Arizona that was almost nonstop birding at well-known "hot spots," such as Cave Creek Canyon, with wake-ups as early as 3:45 a.m.[12] Five men from Baltic, Ohio, harvested four bull elk while tent camping in a snowstorm in Colorado and packed the meat out themselves.[13] Though the quest to add birds to one's life list or to bag a trophy animal lends a quality of single-mindedness to such trips, many birders and hunters become self-taught naturalists by being attentive to the behavior, food supplies, habitat, and microclimates preferred by the animals they are pursuing.

The Roads Less Traveled

Some types of nature-oriented tourism that are sought out by non-Amish do not appeal very much to Amish sensibilities. We encountered virtually no seekers of extreme physical challenge in nature, such as skydivers or rock climbers, with the explicit travel goal of testing the limits of one's physical capacities. Taking risks in nature to push the limits of the body and have a transcendental experience does not resonate with an Amish understanding of how one achieves a "peak experience," through service to God and the community.[14] Another type of nature-oriented travel that remains virtually unknown in Amish circles is ecotourism. For Amish travelers, cost is usually a primary determinant of where and how they travel; they almost never consider the ecological sustainability of the tourist venue when making their choices. In the use of resources, however, Amish practices are consistent with one goal of ecotourism: minimizing consumption. With the exception of the New Order, Amish tourists typically do not fly, they travel in large groups in one vehicle, and they are generally frugal in their choices of accommodation, recreation, and food. One woman summarized the mind-set she had grown up with as follows: "We shouldn't go to that expensive place if we can find a cheaper place." Their modes of travel are thus partially aligned with one of the goals of ecotourism even though the Amish do not usually make a conscious decision to be ecologically minded travelers.

Travel for outdoor recreation is sometimes assumed to be a pro-environmental practice that creates stronger support for the conservation of natural areas. But some research cautions against the notion that nature travel automatically leads to conservation mindedness.[15] Amish who travel on organized tours to natural sites or just see nature as a backdrop may observe the effects of urban sprawl or how green spaces can coexist with urban areas. Those who travel out West often return with a heightened awareness of the impact of water shortages, the danger of forest fires, and the effects of prolonged droughts. Birders and hunters acquire a clearer understanding of increases or declines in populations of sought-after species. But such observations do not automatically lead to deeper understanding of the root causes of changes in ecosystems.

Whether increased awareness of diverse landscapes translates into a newfound commitment to environmentalism is less clear. While travel may lead to a greater appreciation of natural areas and an enhanced conservation ethic, it may also confirm preexisting biases. For example, some Amish travelers reported that spending time in a wilderness area had convinced them of the importance of preserving large tracts of land for future generations. Others became convinced, however, that the government had "locked up" too much land and that spurious environmental concerns like the spotted owl had prevented responsible logging and contributed to the uptick in forest fires. In this respect, the Amish are no different from other tourists who process their experiences through a cultural filter that is only somewhat amenable to modification.

Amish Expression: Emerging Nature Writers and Artists

On November 21, 2015, approximately 150 Amish men and women gathered at a Mennonite church in Berlin, Ohio, for an event that would have been remarkable even twenty years earlier: the Fourth Annual Writers and Artists Meeting. In addition to hymn singing and book presentations by authors, the mostly Old and New Order attendees from Michigan, Ohio, and Indiana listened to talks on topics such as poems and songs, the value of print, and stories brought to life by descriptive word usage. The goal of the meeting, as stated in the program, was "to wake up dormant talent among the horse-n-buggy people" and "to polish active talent for the glory of God, to the building of his kingdom." Workshops such as this one respond to the rising interest in writing and drawing among the Amish, evidenced by the growth in Amish-edited subscription periodicals, which now number more than fifty.[16] In light of this surge in writing and accompanying artwork, the term "Amish fine arts" can no longer be considered an oxymoron.

The entry of the Amish into the publishing world in the mid-twentieth century was tied to the rise in the number of Amish private schools.[17] When Pathway Publishers in Aylmer, Ontario, began to publish parochial-school textbooks and the trio of *Family Life*, *Blackboard Bulletin*, and *Young Companion* magazines, it marked the emergence of an attempt to "intentionally reinforce contemporary Amish identity."[18] In these formative years of Amish publishing, humility was paramount, names of authors were often omitted, and writing had to be, as one Old Order man put it, "working for the good of everybody else." The recent emergence of "Amish expression," however, represents a new cultural development in that individual Amish men and women are not only editing and writing under their own names but also producing and selling a diverse set of works: travel memoirs, cookbooks, how-to manuals, even sketches and paintings.

Most Amish writing is designed for and consumed by an Amish audience. But the cachet of the Amish brand, with its positive association with the values of simplicity, quality workmanship, rurality, and family togetherness, has created a growing, if uneven, climate of receptivity to Amish voices among outsiders. "Can you imagine that suppressed feeling of centuries where you were hated and ostracized, and nobody wanted to hear what you had to say?" reflected a published Old Order author. Inside the Amish community, a different dynamic is at work, as aspiring authors and artists must balance their chosen forms of written expression to make sure they do not seem too worldly to potential readers. Some writers and artists point to entrepreneurs in their community who go to furniture or other trade shows and compete directly with non-Amish to bolster their case. If they're being self-actualized in their businesses, the argument goes, then "that opens it up for us to say, 'This is our trade,'" reflected one well-known Amish author. Tying these forces together at the institutional level are the growing number of publishing houses in all the large settlements that cater to the Amish and actively seek out manuscripts, creating acceptable outlets for Amish writing.[19]

An Amish Toolkit for Expression about Nature

In the rural landscape where Amish live, writing about nature encompasses a dimension of the human experience to which everyone can relate. One Old Order writer noted, "Just below doctrinal writing, nature is second on the list." Asked to elaborate on the qualities of good writing, he said, "Don't take advantage of people, always use true stuff, say it simply, with a little bit of wit, and enter nature lavishly into your writing. Introducing a Bible verse to illustrate what you want to say doesn't hurt either." Even though nature is a relatively "safe" subject compared with politics, powerful cultural currents steer Amish nature writers in certain directions and not others. Most Amish who write about nature draw from a cultural toolkit that includes sharing personal experiences, presenting factual material, and using nature to tell a morality tale and to celebrate God's handiwork.

Autobiography: Personal Experiences in Nature

The most common form of written expression about nature is autobiographical. Drawing on personal experiences has the advantage of being easy for the reader to relate to, but it also shields the author from the possible criticism of claiming too much from a limited knowledge base. One published author confided, "See, I just write about things that I do. I take notes, and I build a story around that. I keep track of the birds I see and just write about things that happen." Many of these firsthand accounts also provide factual information about nature and blend into morality tales. Yet establishing one's personal experience is crucial in achieving a sense of legitimacy when writing for an Amish readership.

Self-published books about adventures in the outdoors abound. Martin and Susan Hochstetler wrote a trilogy, for example, that begins with *Life on the Edge of the Wilderness*, which chronicles "our own story of the two years we lived on the ranch in the Cariboo region, in British Columbia."[20] Chapters on trapping, hunting, and encounters with cougars are interspersed with accounts of other aspects of frontier life, such as calving time and accidents that led to hospital visits. Some Amish-edited periodicals, such as the *Connection*, have regular nature columns with featured writers. Others, like *Hometown Outdoors*, rely on new writers who submit material; the January–February 2016 issue features stories about personal experiences ranging from spring turkey hunting to a day of ice fishing to a coyote hunt. Unlike autobiographical accounts that are deeply reflective about the larger meaning of the chase or wax philosophical about the transformative power of nature, most Amish autobiographical accounts are blow-by-blow descriptions of events and actions.

Factual Writing about Nature

Another acceptable mode of nature writing is sharing factual information that may be new to the reader. Amish volunteer correspondents who write for weekly newspapers that cater to an Amish readership across the country, such as the *Budget* and *Die Botschaft*, regularly include updates on the weather. But nature columnists for popular Amish periodicals often write about particular organisms that readers might encounter in order to convey information about their habits. Writing in a column on birds for the *Connection*, for example, Joe and Rosemary Bontrager introduce readers to the tree swallow. They describe its appearance, range and migration patterns, diet, song, and nesting habits, including an account of how house sparrows compete with tree swallows for nesting sites.[21]

Many Amish authors who convey factual information about the natural world do not bother to acknowledge the source of their material, and it is not uncommon for them to borrow from encyclopedias or nature guidebooks. However, according to one published author in Illinois, facts that come from scientific sources and are not readily observable are almost always presented with a qualifier. He gave three examples: "*scientists claim* that the moon's light is a reflection of the sun's light"; "gold is a heavy metal, *or so they say*"; and "*according to the doctors*, our bodies are three-fourths water." His explanation for this tendency is twofold: humility is valued more than knowledge, and reliance on scientific sources is suspect. Science is seen as shifting sand: the findings of today may be overturned tomorrow. By contrast, he says, "God's word is unchanging, and our own experiences are real."

The artistic equivalent of writing factual material about nature can be found in the growing numbers of Amish who draw, sketch, and paint animals or natural scenes. Many Amish children grow up with coloring books featuring real-life scenes and animals, informally competing with siblings to "color within the lines." It is a logical next step to focus on drawing or painting an animal to realistically approximate its appearance in nature. Many magazines hold regular drawing contests for children and publish the winning drawings in each issue. *Journey of Wings*, a birding magazine targeted at Amish across the country, holds a quarterly bird-drawing competition for children, teenagers, and adults; the drawings must be freehand but can be in pencil, pen, watercolor, or colored pencil. A column by the artist Jamin Schrock in the fall 2016 issue gives advice on how to sketch a bird as realistically as possible. "Start with the eye," advises Shrock. "The main thing here is to get some shine into the eye so it looks alive."[22]

Using Nature to Tell a Morality Tale

Another acceptable form of nature writing is to tie nature to a moral message that is widely affirmed among the Amish, such as celebrating hard work or helping others before taking care of your own needs. The Amish-owned Pathway Publishers, which caters to the moderately conservative Amish, specializes in this type of nature writing in its magazines and in works of fiction for boys and girls aged six to twelve. Books such as *Eli and the Purple Martins* or *Menno's Ducks*, which fall into the category of "realistic fiction about stories that could be true,"[23] teach lessons on responsibility in caring for animals. Nature-themed columns in periodicals frequently have a strong moral undertone as well. The author of a nature column in the children's magazine *Shining for Jesus* first asks why God would create ants, and then, having solicited "interesting facts about ants" from children who read the

BIRD drawing contest

CHILD; 11 YEARS AND UNDER

THREE AGE LEVELS:
Child: 11 years and under
Teenage: 12-17 years
Adult: 18 + years

RULES:
All drawings must be free-hand, but can be in pencil, pen, water color or colored pencil.

PLEASE DO NOT use ruled paper, write on your drawings, or fold them.

If you want your original sketch back, please include a SASE to return, and also a note asking us to return the sketch.

THE WINNERS RECEIVE A PRIZE!
It's up to our readers who will be winners. Please choose the ones you like best of each age level and send your votes to the address below. If there is more than one with the same amount of votes we will draw a name out of the nominees.

The Bird Drawing Contest is not associated with the Sketch a Bird on page 8.

Send all correspondence to:
THE JOURNEY OF WINGS
6954 County Road 77
Millersburg, Ohio 44654

1-C *Red-headed Woodpecker*
2-C *Swallow-tailed Kite*
3-C *Hooded Merganser*
4-C *Hooded Merganser*

Last issue's Bird Drawing Contest winners were:
2-C by *Adam Yoder*, Age 11, Fredericksburg OH
2-T by *Caleb Wenger*, Tunas, MO
3-A by *Mary Miller*, Bremen, IN
The winners received $10.00 gift coupon.
Great job to all of our contestants!

—20— the journey of wings

drawings continued on next page

Many Amish publications feature sections and activities for children, including drawing contests such as this one announced in an Amish-published birding magazine, *Journey of Wings*. Photo by Marilyn Loveless

magazine, weaves a moral message about knowing one's role and working hard as a team.[24] An article in *Family Life* titled "Are You an Invertebrate?" tells readers that "an invertebrate is a mammal with a backbone," then asks, "Do you stick to your convictions even if your friends don't—or are you a jellyfish?"[25]

Artists also use nature as a canvas to convey moral messages. Perhaps no better examples exist than the sketches of horses titled "Pencil Passion," by Andy Mast, an Amish man from southern Illinois. When he was seventeen, Mast suffered a serious head injury while walking a horse to pasture on the family farm, and "through his dark time, drawing became a sanctuary." His purpose is to use his "God-given talent"

Andy Mast's award-winning sketch *A Long Day* is an allegory for his struggle to regain his health after a bout with illness, but it also captures the emotional bond between horse and rider. Courtesy of Amish Pencil Artist Andy Mast

to "bring his viewers courage, peace and hope through his art."[26] Mast specializes in pencil sketches; in this sense, he meets the expectations of most Amish, who, as one woman artist put it, "think of art as a pencil or pen drawing or a painting." He also represents a long tradition of Amish with disabilities, who work as artists for financial gain in ways that go beyond the bounds otherwise allowed for individual expression. By his own admission, however, Mast pushes the limits of acceptability in the way he has marketed his work. Mast entered and placed first, for example, in the Western Spirit Art Show and Sale in Cheyenne, Wyoming, and he has designed a professional advertising brochure that mentions his Amish identity. "Some people would frown on it because he's trying to build a little kingdom, or fame," commented one author. The way Mast weaves together his personal story, his talent for drawing, and a savvy marketing plan illustrates how emerging Amish authors and artists can tie a nature theme to an uplifting personal and religious message and gain legitimacy and sales among segments of the Amish community and beyond.[27]

Amish poets frequently celebrate natural beauty or natural occurrences as a sign of God's omnipotence. "If someone needs a subject for poetry, it's often nature," commented one author. Amish poets usually find sources of inspiration in daily life. One man we spoke with started writing poetry when he was milking cows and plowing the fields. "My sketch pad was my arm all the way back to my elbow," he said, "and if it got too sweaty, I would write on my shirt—until I met too much opposition from my wife." Largely because of familiarity, Amish poems are usually written in iambic pentameter. "I would've grown up thinking a poem has four lines and it rhymes. It's just rhyme prose," reflected one published poet. The careful observation of nature in one's own life, interpreted as a reflection of God's handiwork, is on display in the following poem, titled "His World."

The waiting hush before a storm—
The pitter-pat of rain
Drumming to the beat of love
On our windowpane
Close your eyes and listen
To the brush of leaves unfurled—
Sense the majesty of God
The heartbeat of His world.[28]

Writing on Thin Ice

In the Amish world-view, not all ways of framing and writing about nature are benign. The Amish believe that God transcends nature and "created all things visible and invisible," which he governs by his "wisdom, might and the word of his power."[29] Several implications for acceptable nature writing follow from this biblical foundation. Any mention of spiritual forces that are embedded in nature or that sound too much like New Age philosophy are suspect. "I wouldn't put in too much about meditating and Eastern religions," one Old Order author advised. Depicting nature as pulsing with energy or some other life force can also be problematic. An Old Order author cautioned that writing "Mother Earth" in capital letters should be avoided because it "gets too close to nature worship." Moreover, giving human actions or thoughts to animals is strongly discouraged because it is seen as blurring the line, ordained by God, between humans, who have a soul, and other creatures, who do not. Collectively, these shared understandings about what's appropriate lead Amish nature writers to keep fictional accounts closely based on a "true story" and to avoid genres such as fantasy and science fiction. To be credible within their community, writing must pass the test of verisimilitude.

Even though nature is widely seen as evidence of God's handiwork, delving too deeply into the subject can paradoxically "raise eyebrows" in the Amish community. Going into great depth about the natural world also brings an author closer to scientific modes of writing and, by implication, to another highly charged topic. "If you write about evolution, that's an absolute no-no," said one author. The sensitivity of this issue was revealed in reactions to a 2013 article in *Farming* magazine in which a non-Amish contributor wrote, "I was taught by a man who thought highly enough of us to explain that there is no conflict between Darwin's theory of evolution and the Genesis account."[30] In a letter to the editor, an Amish reader wrote that he was "deeply disturbed" by the article because "the false theory of evolution has done more to lower our nations high moral values, and bring all its accompanying evils, than any other thing." His and other readers' calls to more effectively screen magazine content prompted the editor to write, "The idea that Darwinism is compatible with Genesis is not shared by the editors and publishers of *Farming*, it is the author's own view."[31] Though incidents such as these are rare, they nonetheless send a powerful signal to Amish authors about where the boundaries for acceptable nature writing lie in their community.

An Environmentalist Message: The Writing of David Kline

In the introduction to his anthology *American Earth*, the environmentalist Bill McKibben asserts that environmental writing might be "America's single most distinctive contribution to the world's literature."[32] He goes on to argue that environmental writing is different from nature writing in that it moves beyond knowledge of and appreciation for the natural world to ask searching questions about what happens when people collide with nature. In other words, environmental writing recognizes that the natural world is no longer invulnerable, and it connects descriptions of nature to larger moral, philosophical, or political issues, opening the reader up to a wider universe of possibilities. Even though many of the best-known American environmental writers were deeply influenced by their Christian beliefs,[33] the vast majority of Amish who write about nature rarely fashion an explicit environmentalist message.

There is one noteworthy exception, however. As we spoke with Amish across the country about their views of nature, many were aware of the works of one organic farmer in Ohio: David Kline. Among non-Amish naturalists and environmentalists too, Kline's writings are well known for their eloquent and passionate defense of small-scale farming and stewardship of the natural world. Kline has published three books, all collections of short essays on the virtues of agrarian life. His message has been celebrated in books by Wendell Berry, Bill McKibben, and Barbara Kingsolver, among others. One of Kline's most recent literary projects has been editing *Farming* magazine, which, according to its mission statement, "celebrates the joys of farming well and living well on a small and ecologically conscious scale." Kline still writes a short essay for each issue of the magazine, more than half of whose three thousand subscribers are non-Amish.

Kline's writings include all the elements of conventional Amish narratives about nature mentioned above, but they also go beyond them in several important ways. For one, his work is informed by other naturalist-writers, and he connects his essays to the larger genre of environmental writing. In an essay on weeds, for example, he quotes Ralph Waldo Emerson, who wrote that "a weed is a plant whose virtues have not yet been discovered."[34] Writing about wildflowers and ferns, he references Aldo Leopold's statement that "there are some who can live without wild things, and some who cannot" and places himself, and by extension the reader, in the latter category.[35] He reminds us of Rachel Carson's observation about the "sense of wonder" connected to nature.[36] And the list goes on. Barry Lopez, Edwin Way Teale, and even poets such as Edgar Allan Poe and William Cullen Bryant are appropriated in the service of his message. Kline's willingness to demonstrate the breadth of his reading and his ability to use a specific observation in nature to open up a larger window on the world show similarities with the best of non-Amish environmental writing.

Kline is distinctive among Amish nature writers in his tendency to point out human actions that have adversely affected other organisms and to cite scientific evidence. In an essay on the gray tree frog, for example, Kline notes that frogs are vanishing worldwide and scientists aren't sure why but that possible reasons include the pollution of wetlands and depletion of the ozone layer.[37] Against the cultural grain, he defends the ecological role of snakes and dismisses as superstition and myth the fear of snakes in rural areas,[38] never mentioning the Amish by name. In one particularly memorable passage, he recounts that while he was cutting firewood in a woodlot of dead and diseased trees, "a voice whispers in protest, 'Woodman, spare that tree.' It is my inner voice, my conscience, speaking for the creatures of the woods that depend on dead trees for survival."[39] Here Kline asks us to consider nature from the perspective not of humans but of other living beings in the forest. Underlying all his works is a plea for wise stewardship and a critique of the "dominion" interpretation of Genesis: "Does this give us the right," he asks, "to exploit, abuse, and exterminate His Creation?"[40]

Perhaps the most distinctive aspect of Kline's environmental writing is that he is not averse to wading into politics. He frequently takes stands on issues such as hydraulic fracturing and global warming that do not represent the majority views in his rural community. At a time when many Amish were leasing their land to natural-gas companies, Kline questioned the hype generated by the media and gas companies and opined that "the promise of shale gas as an abundant and cheap source of energy for the next 100 years has been wildly oversold."[41] Writing against a deep skepticism in his community about the science behind climate change, he made a series of personal observations, such as that dandelions were blooming in December, and then wrote, "Perhaps Wes Jackson is correct in suggesting that before long strawberries will be grown on Baffin Island."[42] And after mentioning Hurricane Ivan in one issue, he asked, "Are these powerful hurricanes a result of the earth heating up, which in turn spurs more violent storms?"[43]

Kline also calls attention to local environmental issues, such as the health of the Sugarcreek watershed in his own northeastern Ohio neighborhood. Having carefully observed the slow degradation of the spring-fed creek near his farm for more than half a century, he bemoans the loss of fish, eels, and mussels. Then asking the hard question "What has caused the decline of life in the creek?," he points to channelization upstream, changing agricultural practices, including grazing livestock on streambanks, and pesticide use.[44] In another issue, he shines a spotlight on littering and rampant consumerism, describing how he discovered 168 different brand-name items in bags of litter he picked up along two miles of ditches.[45] Buying local is another favorite cause Kline supports, and he often extols the virtues of keeping worn-out bills circulating within the community.

Though many non-Amish readers assume that Kline's writing is representative of the Amish community, it is rare for Amish writers to venture into the territory of environmental writing that Kline inhabits. Kline has had a significant influence on the small but growing Amish organic movement, but the environmentalist strain in his writing is, as one non-Amish observer put it, "a voice in the wilderness." There are several reasons why Amish nature writers other than Kline are reluctant to fashion an explicitly environmental message. One reason is a general feeling that the Amish are already closer to nature than most non-Amish. "We don't feel we should be passing on that message because we're not aware that we're abusing [the environment]," commented one Old Order author. Others acknowledge that the Amish aren't always environmentally minded and locate the cause in the notion that "in the end times, all of this will disappear," according to one Amish author. A female nature writer concurred: "If you write a conservation message, some will think, 'Nature will take care of itself.'" She also pointed out that Amish nature writers have not been exposed to much environmental writing and worry that they will be criticized for their ignorance if they write about a movement they identify as part of the non-Amish world. The absence of a clear environmentalist message in most Amish writing about nature, however, is not at all unusual when viewed in the context of the dominant narrative in rural areas across the United States, where nature is seen largely as a resource and a recreational outlet rather than a fragile ecosystem imperiled by human actions.[46]

A New Awakening

If the goal of the annual meeting for Amish writers and artists is to "wake up dormant talent," then the question naturally arises, toward what end? A similar question might be asked of the swelling ranks of Amish who count themselves as travelers and vacationers. In both cases, Amish across the country are waking up to new possibilities for experiencing the world and new forms for conveying those experiences. For conservative Amish groups, the dangers of this trend seem obvious. New kinds of travel and written expression illustrate growing differences in wealth and increasing reliance on questionable technologies, which will hasten the slide down the slippery slope to individualism and self-actualization. For others, however, these new opportunities provide a breath of fresh air, an opportunity to expand horizons that have been historically and culturally limited. Columnists for Amish periodicals brush up against the ideas of a diverse group of other Amish (and sometimes non-Amish) writers on a regular basis. The social worlds of Amish who can take advantage of these opportunities are thus expanding in unpredictable ways. In Part IV, we further explore Amish responses to local and global environmental issues and ask how those responses illustrate their views of the relation between humans and the natural world.

IV *The Amish as Environmentalists*

CHAPTER TEN

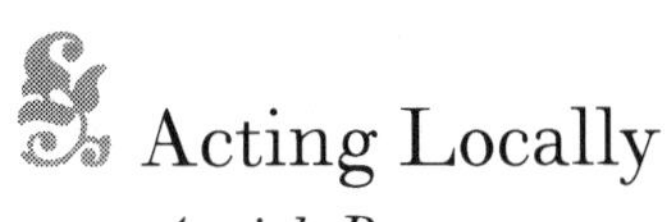

Acting Locally

Amish Responses to Regional Environmental Issues

Amish homes, farms, shops, and communities are embedded within a regional landscape and a political system confronted by a variety of environmental challenges, including air and water pollution, habitat destruction, and loss of biodiversity. Local, state, and federal regulations have been created to address many of these problems, which are shared by Amish and non-Amish alike. The logic behind this regulatory intervention was summarized by the biologist Garrett Hardin in his widely cited 1968 article "The Tragedy of the Commons." Because self-interested individuals or corporations cannot be assumed to act on behalf of the public good when it comes to shared resources such as water or air, Hardin argues, some form of external oversight and coordination is necessary to ensure that resources remain for future generations.[1]

Environmental laws and the agencies that enforce them are relatively recent creations, however. The Environmental Protection Agency and the Clean Water Act, both adopted under the Nixon administration, in the early 1970s, are less than fifty years old. In many ways, the country as a whole is still sorting out and evaluating the kind and degree of regulatory apparatus needed to protect our shared environment. In recent decades, Democrats and Republicans have taken starkly different positions on environmental regulation and its relation to economic growth, while a slight majority of Americans say stricter environmental regulations are worth the cost.[2] We were curious to know how the Amish respond to local and regional environmental challenges and how they think about the idea of ecological limits to human activities in this politically divisive climate. Beyond their response to environmental regulations, we also wondered whether Amish consider it their moral responsibility to protect the local environment and to what degree they identify and acknowledge their role in altering the region in which they live. In this chapter, we consider how Amish ways of life intersect with environmental issues that are manifest within their local or regional communities.

In important respects, their culture and history place the Amish in a position to effectively address local environmental concerns. Their foundational ethic of stewardship of God's creation can be interpreted as requiring respect for all living organisms. As one man put it, "When God created things, he said they were good, so how can we say otherwise?" Their close ties to the land, cultivated over generations, provide a reservoir of local knowledge and may make them keen observers of changes in the natural world. Their emphasis on community and their ability to mobilize church members to raise funds for medical bills and other causes provide a model that could be harnessed to address local environmental problems. Moreover, the Amish have proved surprisingly adept at working with officials at every level of government, from the National Amish Steering Committee's work on alternative military service at the federal level to state- and settlement-level committees that address issues ranging from road repair to parochial-school standards.[3] These social and organizational resources provide a ready toolkit to draw on in addressing environmental concerns.

In other ways, however, well-established Amish patterns of interacting with "the world" may preclude an effective response to local environmental issues. The Amish are widely regarded as apolitical, keeping government at a distance.[4] While they abide by the law and appreciate the legal principles defending the practice of their religion, the Amish engage the state more as subjects than as citizens.[5] As Donald Kraybill reminds us, citizenship demands acceptance of "a modicum of responsibility for the welfare of the larger social order," yet "the Amish are supremely oriented toward internal obligations within their own religious community."[6] This doesn't mean that they are unaware of local issues, but they seldom participate in public meetings, except those on subjects that most directly affect their lives. The Amish share with other political conservatives a deep suspicion of government bureaucrats and the scientists whose knowledge serves as the basis for regulatory intervention. They may see environmental regulations more as a nuisance and a constraint on their activities than as promoting the common good.

These competing factors suggest that understanding Amish responses to local and regional environmental issues is not a matter of simple extrapolation from the contradictory clichés of the nature-loving and antiregulatory Amish. Instead, we must look at the details of Amish engagement with particular environmental issues in specific geographic and temporal contexts to understand how Amish responses make sense within Amish value systems and how the Amish resist, deflect, or work with the state. We turn now to three environmental issues that offer diverse insights into how the Amish think about and respond to environmental issues in their own backyard: farm runoff, hydraulic fracturing, and waste disposal and recycling.

Watershed Moments

With the Amish population scattered across rural areas in more than thirty states, their farms, and especially their livestock, impact thousands of watersheds. As early as 1994, the attorney Elizabeth Place pointed out that since Amish farmers "apply manure on the basis of disposal needs rather than crop nutrient needs, the potential for runoff is great."[7] In recent decades, farms in the largest settlements have attracted increased scrutiny from regulatory agencies. In northern Indiana, the Little Elkhart River Watershed has been the focus of sustained remediation efforts. The Sugar Creek Watershed, which drains much of the Holmes County settlement, in northeastern Ohio, was identified in a 1998 Ohio Environmental Protection Agency study as the second most polluted watershed in the state. In Lancaster, Pennsylvania, Amish farms have been singled out by the high-profile Chesapeake Bay Program, a $1 billion federal effort to reverse the decline in water quality and aquatic life in the bay. An examination of how Amish farmers have cooperated with and resisted remediation efforts in these specific situations reveals much about the interplay of religious ethics, economic interests, and ecological priorities.

Over the past decades, governmental agencies and environmentalists have developed a highly scientific and technical approach to watershed management. Under the Clean Water Act, states must establish Total Maximum Daily Loads (TMDLs), which are the maximum amounts of pollutants bodies of water can contain and still meet water-quality standards.[8] For the three watersheds with high numbers of Amish farms mentioned above, EPA studies have shown fairly consistent indicators of degraded water quality, including include high levels of total phosphorous and nitrates in the water, excessive sedimentation, and levels of *E. coli* that exceed the human health standard.[9] High concentrations of toxic contaminants such as herbicides, PCBs, and mercury have also been measured.[10] The tangible effects of degraded water make headlines when green algae blooms impact recreational water sports or people fall ill from ingesting water with high fecal-coliform counts. Less visible but perhaps of greater long-term concern are effects on the diversity and numbers of fish and other aquatic species and chemical residues in the food chain. The EPA has cited nonpoint sources of pollution from urban, suburban, and agricultural land transported by precipitation and runoff as the leading cause of water impairment in the United States.[11]

Agricultural conditions that contribute to watershed pollution are not exclusive to Amish farms. These include barnyards perched on hills above streams, the lack of buffer zones along banks of waterways, nutrient and chemical runoff from fertilizers

Dairy farmers whose barns are perched above creeks are prime candidates for cost-sharing plans to reduce runoff of cow manure and nutrients, which pollute local watersheds. Photo by David McConnell

and pesticides, and sediment erosion from plowing sloped fields. Because of their small herds and minimal manure-storage capacity, however, Amish farmers are more likely to allow livestock in creeks and to engage in improper manure management. Watersheds with residents from ultraconservative Amish groups sometimes include homes with septic systems that are "straight-piped" into a creek or connected to field drainage tiles. However, establishing causation and responsibility is "notoriously difficult when addressing nonpoint source pollution problems."[12] The assertion that even minor problems in one field's drainage ditch can seriously impact the lower reaches of a watershed is often a point of contention.[13] A major challenge regulatory agencies face, then, is how to persuade Amish farmers that their actions upstream are partially responsible for watershed pollution downstream.

The logic and the language used in watershed management can be foreign to the world-view and lived realities of many farmers, including the Amish. Like many other farmers, the Amish value their autonomy. They see their Christian values, based on hard work and firsthand experience living on the land, as being threatened by a secular, hostile society. Their first reaction is to interpret a technical assessment made by scientists and bureaucrats about their role in downstream pollution as unjustifiable finger-pointing.[14] Even if farmers can be convinced of their role in contributing to watershed pollution, persuading them to adopt best management practices (BMPs) is not easy, especially when there are significant costs associated with the remediation plans. Moving a barn that is perched on a hill above a stream or installing a manure pit that can hold waste for up to six months can easily cost upwards of one hundred thousand dollars. A closer look at clean-water programs in the Little Elkhart, Sugar Creek, and Chesapeake Bay watersheds shows not only resistance to change by Amish farmers but also some surprising examples of cooperation with local government offices.

Combatting Agricultural Runoff in Indiana, Ohio, and Pennsylvania

Realizing the importance of water quality in Lake Michigan and inland rivers and lakes for recreation and health, the Indiana General Assembly in 1988 created the St. Joseph River Basin Commission. The commission began working with the LaGrange County Soil and Water Conservation District (SWCD) to remediate the Little Elkhart River and one of its tributaries, Emma Creek. In approaching Amish farmers, however, the government's usual leverage was missing: it could not withhold farm subsidies, because Amish farmers had not accepted them in the first place.[15] After it "became evident that Amish residents were reluctant to voice opinions in public," the LaGrange County SWCD took the unusual step of asking an Amish man to serve as a board member and on the steering committee.[16] "I'm probably the only one in the U.S. [in this kind of position]," the man told us, adding that seeing the decline in water quality near Lake Emma had gotten him interested in volunteering.[17] Getting buy-in from the Amish community was a matter of finding "a few guys that are good working with people, who take a soft approach, no arm-twisting." He said it was relatively easy "once we got the bishops on board."

The combination of efforts to place Amish in key committees and an aggressive but soft outreach campaign was instrumental in persuading the Amish community to accept governmental funds to partially or wholly cover the expense of remediation.[18] Key projects included laying water lines to the back pastures, putting in fencing to keep cows out of the stream, and installing water-testing stations on the creeks to provide regular data on water quality. The willingness of the Elkhart-LaGrange Amish community to engage with governmental agencies was essential to their participation in cleanup programs.[19] Amish attitudes toward cost-sharing "have changed a lot—25 or 30 years ago, that would've been a no-go," said the Amish board member. "Farmers have gotten really good at networking and adopting best practices because they see that it all comes down to making a living." Though many challenges remain, in 2011 the EPA touted reductions in livestock-induced pollution in Emma Creek as a success story, noting that landowners independently paid $30,000 of the $2.3 million price tag for remediation of the watershed.[20] When Amish are willing to take on leadership and liaison roles to promote appropriately designed governmental programs, farmers may be more willing to adopt BMPs.

In contrast, efforts to improve water quality in the Sugar Creek Watershed, which drains eastern portions of the Holmes County settlement in Ohio, did not lead to action by the Ohio legislature, even though a 2002 study by the Ohio Environmental Protection Agency found the watershed second only to the infamous Cuyahoga River in impairment.[21] Instead, a group of researchers at Ohio State University's Agricultural Research and Development Center received several grants to form "a community-based approach to watershed management that emphasizes local action and decision-making based on scientific data."[22] The effort had many successes and persuaded some Amish farmers to improve their agricultural practices. Without political backing from the state legislature or congressional representatives, however, and because the Holmes County settlement includes a range of liberal and conservative Amish groups, it proved difficult to coordinate responses across the watershed.[23] Our conversations with Amish farmers confirmed the findings of an innovative "farmer learning circle" convened by the Ohio State University researchers: local farmers were deeply skeptical about the methodology and conclusions of the 1998 study.[24] In a 2010 study of the Swartzentruber Amish community in the most severely impaired area of the Sugar Creek watershed, "none of those interviewed gave any inclination that they were aware of the polluted conditions of the Sugar Creek Watershed."[25]

As in northern Indiana, bridging the perception gap between the world-view of Amish farmers and science-based environmentalism has fallen to the county SWCDs. As the primary link between farmers and the government, SWCD personnel can greatly influence whether farmers adopt conservation agriculture. SWCD offices sometimes implicitly support intensive agricultural production methods that exacerbate soil loss and degrade water conditions.[26] However, when SWCD staff are experienced farmers who promote the message "that conservation is good for production," they can gain the confidence of Amish farmers because "we can go out and talk their language and know what they're dealing with on a day to day basis." In one case touted as a national model for nutrient-trading programs, SWCD staff in Holmes and Wayne Counties persuaded Amish farmers to accept funds from a local cheese company to redirect milk-house drains or keep water off feedlots.[27] Their strategy? "In every neighborhood, there is somebody that the Amish look up to. We would go to him and say, 'Look, would you be willing to host a neighborhood meeting if we come and present this?' That model worked very well." In their promotion of nutrient management and other conservation measures, SWCD staff continually hammer home the message that "it's not a matter of if we're going to have [regulations], it's a matter of time."

Judging by media coverage of the EPA crackdown on Amish farmers in Lancaster County, Pennsylvania, that time of reckoning has arrived. A federally coordinated effort to clean up Chesapeake Bay determined that agriculture in Lancaster County, which generates six times as much manure as neighboring counties, was contributing substantially to the pollution of the bay.[28] In 2009 the EPA director and other officials swooped into Watson Run, a pocket of Lancaster County identified as one of three "hot spots" polluting the Chesapeake Bay Watershed. They found that 85 percent of farms were in violation of the law requiring nutrient-management plans.[29] Seventeen were found to be managing their manure inadequately and thus contributing to degraded water quality. Six of the nineteen wells sampled also contained dangerous levels of *E. coli*, and sixteen had nitrate levels exceeding those allowed by the EPA.[30] The inspection attracted the attention of major media outlets and seemed to confirm the suspicions of environmentalists that the Amish were finally being subjected to the scrutiny they deserved. In the wake of this and other ongoing disputes over land management, the historian Steven Nolt observed that environmental concerns have become one of the biggest sources of tension between Amish and non-Amish residents in south-central Pennsylvania.[31]

Direct federal involvement in Lancaster County certainly got the attention of Amish farmers, but suspicions remained about the EPA's motives. "I don't trust those guys," confessed one dairy farmer as we sat on bales of hay in his barn near Gordonville. Fear of losing productivity is the most common reaction among Amish farmers, according to a local non-Amish farmer. "They're afraid they'll have to give up land. There's always some kind of a setback." A closer look at the situation in Lancaster County reveals, however, that significant efforts to work with Amish farmers through the local conservation district had been under way long before the Chesapeake Bay program—and continue to this day.[32] One conservation-district official tasked with outreach to Amish farmers reflected, "It helps to have Anabaptist roots and be a farmer. I don't like to talk about the Bay. I haven't mentioned it for 3–4 years. I focus on what's good for their farm." As a result of the intense scrutiny of Watson Run, all twenty-four farmers drew up management plans and adopted BMPs tailored to their farms. But the countywide response has been mixed. "Some really get it. You know, they want to do it. They understand the whole ecological thing," commented another outreach coordinator. "Then there's some other ones that I don't think get it." He estimated that about half of the Amish farmers in the county had drawn up management plans and noted that the best way to get conservation plans on the farms is for the township to make it a requirement for obtaining a building permit.

Once a conservation plan is drawn up, persuading local Amish farmers to accept money from the government can be challenging. The Lancaster Amish settlement is quite progressive in terms of farm technology but somewhat conservative theologically, which makes it difficult to sell cost-sharing programs across the board. Still, Amish responses vary. "Oh my, yes, we have Amish who will take anything they can get to those who won't take anything," said one conservation-district employee. "The church is never going to endorse it and say, 'Go get money,' but in many cases, they won't say anything against it." Some Amish farmers have participated in the Conservation Reserve Enhancement Program (CREP), which pays farmers three hundred dollars per acre to plant and maintain a thirty-five-foot-wide forested buffer around waterways. The poster boy for cost-sharing, though, is an Amish man in the Churchtown area who did everything the conservation district suggested. Said a district staff member, "I can take somebody there, and I can show him eight different best management practices. And we've been doing a lot of work in that neighborhood since his project." This example demonstrates how one or two individuals can be catalysts for change in a community. But the farmer admitted that it was topsoil loss from his own farm that had finally motivated him to take action. This is a reminder of how unusual it is for Amish to think about the larger public good. "What happens downstream, I never gave a thought," he readily admits. "If they would have said, 'You're going to help save the bay,' I would have said, 'Big deal.' Yet, it all ends up down there; that's true. But it's too far away. That's not what people here are thinking about."[33]

One crop grown by many Lancaster-area Amish farmers that poses real environmental challenges is tobacco. Though the public-health campaign against smoking has made tobacco growing an ethically questionable practice in some circles, it is still a powerful and positive symbol of conservativism for a minority of Amish in the Lancaster community. "Other farms don't do tobacco because it's too much hard work," said one grower. A mother of eight whose family has nine of their one hundred acres planted in tobacco told us, "It makes a big, good winter job for the

Tobacco has made a resurgence among some Amish farmers in Lancaster, Pennsylvania, where buyers are encouraging them to grow high-end specialty varieties for a niche market. Photo by Doyle Yoder

whole family. Even the six- and eight-year-olds can strip the tobacco." Banks promote tobacco when giving out farm loans because it is an economically dependable crop grown on contract with Phillip Morris or other companies.[34] But it requires a lot of sprays and frequent cultivation, disturbing the soil and increasing soil erosion. To help minimize soil loss, the Lancaster conservation district worked with an Amish farm-equipment business to develop a horse-drawn, no-till tobacco planter, which they are currently promoting to Amish tobacco farmers.

As a result of concerted action in Pennsylvania and other states, the health of the Chesapeake Bay is slowly improving.[35] A 2016 news release from the Chesapeake Bay Foundation, however, singled out Pennsylvania as lagging behind other states in reducing nitrogen runoff and asked for an additional $20 million from the USDA and federal partners, with Lancaster County as a major recipient of funds.[36] Whether this economic enticement will be effective in reducing nitrogen runoff is still an open question. Many Amish remain vocally skeptical of what they see as "the environmental agenda." Although Amish farmers seldom engage in political activism, it is no accident that many belong to the American Farm Bureau Federation, a powerful lobby that has fought the Chesapeake Clean Water Blueprint all the way to the Supreme Court. Yet examples of Amish farmers adopting BMPs suggest that the calling to be stewards of the land still resonates. One farmer said that when it rained, "it flushed my barnyard down to the stream, and it kind of bothered me."[37] As we have seen, Amish farmers are willing to work with local and regional agencies if they are approached in the right way. The Amish board member in the LaGrange County SWCD put it this way, "Rather than say, 'Get your cows out of the ditch,' let's take a soft approach and work together to solve the problem. It's a cooperative deal."

The Amish and Fracking

In June 2013, as the hydraulic fracturing boom centered on the Utica and Marcellus shale formations was spreading across western Pennsylvania and eastern Ohio, the *New Republic* published an article with the provocative title "The Amish Are Getting Fracked." The article profiled the case of an Ohio Amish couple who signed a contract with Kenoil allowing the company to lease 158 acres for ten dollars per acre, with the right to drill for shale gas on the property, only to discover that other landowners were being offered upwards of one thousand dollars per acre and a signing bonus. The article argued that oil and gas companies were taking advantage of the Amish because "their religion prohibits lawsuits—and the energy companies know it."[38] The Amish have a long history of leasing their land to energy companies, for strip mining of coal, for natural gas, and now for shale gas. Some of those companies have no doubt played fast and loose with farmers they considered easy targets in the free-for-all competition for drilling rights.[39] Focusing narrowly on how companies take advantage of some Amish landowners, however, misses the broader question of how the Amish themselves respond to fracking from social, economic, and environmental perspectives.

The Amish view that God put everything on the earth for humans to use applies as much to underground resources as to aboveground ones. "If you have a crop of corn, you'll harvest it, so why not harvest the gas?" asked one New Order man. Along with many other rural residents, Amish landowners in southern New York, northeastern and southwestern Pennsylvania, and southeastern Ohio in 2010 and 2011 were caught up in the excitement of a modern-day "land rush" as oilmen from Chesapeake and other energy companies went door-to-door offering leases for thousands of dollars per acre, as well as signing bonuses and a percentage of royalties should a well be dug on their property.[40] Landowner associations formed

Amish landowners in some settlements have a long history of leasing their land for mineral rights and have few concerns about the ecological consequences of new techniques such as hydraulic fracturing. Photo by Doyle Yoder

almost overnight to pool properties and negotiate a better deal with the energy companies, and stories of farmers cashing in spread like wildfire. "One guy got a one-time check for $175,000. That would be nice. It's hard to walk away from that," said one Amish farmer. In the end, most who signed leases did not experience drilling on their property or qualify for royalty payments, because only certain areas proved productive for drilling. The Amish conversations around fracking are worth listening in on, however, because they reveal a preoccupation with the social and economic implications of fracking rather than its environmental impacts. One Old Order man summed up the concerns this way: "Will the prosperity hurt us, will it increase traffic, will it increase the price of land, and could it put an end to tourism? I have yet to hear anyone express environmental concerns."

We frequently heard concerns expressed about congestion and increased truck traffic, especially if a well was located in an area with heavy buggy and bicycle use. "How can a small Amish community withstand the influence of 50 trucks a day going down a gravel road? The horse and buggies, where would they be in all this?" asked one man. Noise pollution was also a concern. "Do you think this is the last quiet fall we'll have in [our] county?" wondered one Ohio schoolteacher. "I went to southeast Ohio, and I saw how it ruined the Amish community," reflected one Old Order man. "It looked like a construction site." Beyond the noise and the congestion, the potential disruption of an influx of workers from Texas and other areas was also a concern. "People come in from other cultures, too, you know, and some of the people live in the oilfields and that's about all they do. . . . Yeah, I was apprehensive."

The concern about fracking's potential to exacerbate the gap between the haves and the have-nots resonated strongly within the Amish community. "I'm personally more concerned about its impact on the structure of our community," said one schoolteacher. "If I have five acres and he has 120 acres, he's getting millions, and I'm getting thousands." An Amish man who served on his town's planning commission commented, "Amish are no different than anybody else. The power of big money can bring spiritual corruption."[41] Some Amish landowners have even retained the mineral rights when selling their land, a practice that was frowned on in the past. Since shale-gas leases are not recorded differently from natural-gas leases, it is impossible to know what percentage of Amish landowners who were approached by shale-gas companies actually signed them. "I think some of my neighbors signed, they haven't come out on that yet," mused a Swartzentruber lumberyard owner. "But some of them are joining landowners associations." Other Amish opinion leaders were more confident in predicting Amish interest. "Nine out of ten will sign," said one Old Order farmer. "When it comes to the almighty dollar, the Amish yield."

After the first round of enthusiasm, the Marcellus and Utica "shale play" faded, though new technologies may revive it in the future. By 2018 only a handful of Amish landowners in the region had received significant lease money or royalty payments. Some individual landowners kept the money and used it to pay off debts, make new investments, or help family members. In Greenville, Pennsylvania, where two wells were dug on Amish properties, the landowners took a different approach. "They sat down as a church, and they decided that the money would go into a central fund," said one New Order man, "not only to the landowner, but to be used for medical bills and schools and stuff like that, and give it away to other communities."

The impact of fracking on water quality came up only infrequently. "The problem of pollution of the aquifers for the future and the vast amount of water it takes to pump the wells—those are real concerns," reflected a New Order man. More often, Amish residents dismissed concerns about water quality as being greatly overblown. One man joked, "For the kind of royalties they pay, I could buy a lot of bottled water!" An Old Order minister used an analogy to make his point: "We have millions of cars on the road a day and it's okay to have a hundred thousand accidents a day, and people lose their life over an automobile, but then we have an absolute hissy fit about one little water source that got damaged by fracking. So let's put it in perspective." Many other Amish we spoke to were equally unconcerned. "There's not much danger of pollution because these are 8000 feet wells, not 1000 feet. The injection wells are more of a problem," said an Old Order sawmill operator.

Aside from their emphasis on income inequality, Amish narratives of complaint and benefit are not that dissimilar from those heard in other rural communities affected by fracking.[42] Overall, the Amish living in areas favorable to fracking are more concerned about its potential to upset the quiet environment of rurality than its threat to the water supply. "A lot of Amish people believe that there's a set amount of water that was put on the earth," said an Old Order man, "and because it keeps recycling, the idea that there could be a water shortage globally seems kind of ridiculous." This assumption that water is an unlimited, God-given resource is still widespread, although Amish who have moved to Colorado and Montana have discovered that negotiating water rights is even more important than negotiating mineral rights when buying property.

Waste Disposal and Recycling

While the opportunity to lease land for fracking avails itself to only a small number of landowners perhaps once in a lifetime, disposing of garbage and used goods is a task that ordinary people face every day. What we count as waste and how we dispose of it are affected by the institutional structures and opportunities around us. Our decisions also say a lot about our environmental consciousness, since the monetary rewards for ecologically sound waste disposal are often minimal. Noting that the average American citizen generates 102 tons of garbage over a lifetime, the Pulitzer Prize–winning author Edward Humes argues that garbage has become "one of the most accurate measures of prosperity in twenty-first century America and the world."[43] The "throw-away society" we live in also speaks volumes about our relationship to the environment, as revealed by the mounting evidence of harmful effects of plastics in our lakes and oceans and the contamination of soil, water, and atmosphere from landfills.

We explored Amish sensibilities and practices in terms of the 3 Rs of solid waste disposal—reduce, reuse, recycle. As discussed in chapter 3, the Amish stand out in certain ways for reducing consumption, growing and buying food locally, and limiting material purchases, especially of electronics and transportation. Gift-giving around birthdays and Christmas tends to be restrained and to focus on practical items. These behaviors do not stem from an environmentalist philosophy nor even from a religious vow of asceticism; rather, they are tied to "the sociological and spiritual needs of the community."[44] At the same time, signs of growing affluence can be seen in many settlements. More liberal Amish homes often have up-to-date amenities and include children's play equipment, decks, grills, fire pits, and lawn furniture. Observing the discrepancy between Amish and non-Amish families in standard of living, one church member in St. Ignatius, Montana, lamented, "Sometimes I think we should live in more humble houses. It's hard to be zealous Christians when we are so prosperous."

Amish values of thriftiness and frugality also lead them to reuse durable goods, like clothing and furniture, for as long as possible. Swartzentruber Amish are legendary for wearing clothes to within an inch of their lives, while hand-me-downs are common in most extended families. The Amish frequently make both business and household purchases at secondhand markets. "Goodwilling," or bargain shopping for clothes and other items at thrift stores and yard sales, is a favorite activity of many Amish women, and Amish-run bent-and-dent grocery stores are a staple in most large communities. Some Amish businesses have even learned to capitalize on new markets for recycled materials. We visited a Shipshewana, Indiana, Amish business that advertises birdhouses and bird feeders made from "Eco-Friendly Polywood," a material the owner said comes from recycled milk cartons. The prospects for reusing sawdust from lumber mills have also improved greatly because it can be sold for eco-bricks, animal bedding, and other uses.

When it comes to the third pillar of the triad, recycling, our survey results call into question a *Business Insider* report titled "13 Money Secrets From The Amish," which claimed that the Amish "take recycling to 'unbelievable levels.'"[45] Some Amish did tell us that they believed recycling was important. "No, we shouldn't waste resources. We should recycle. That's good," commented one man. But most observed that it was not a point of emphasis in the Amish community. "It's not a topic I hear discussed a lot," confided one man. "I don't see much of that, recycling," said an Old Order businessman. "To get into recycling big-time, you have to think long-term." As table 10.1 shows, Amish respondents were less likely than their rural non-Amish counterparts to say they recycled most common materials. Since aluminum and metal can be sold to recycling outlets, Amish were more likely to recycle these materials. But for things like plastic, glass, and paper, recycling rates among Amish were quite low. The Amish use 12-volt batteries in many applications, and those too can be returned for cash. One Old Order man observed, "I know our family

TABLE 10.1.
Percentage of respondents to our survey who reported recycling various materials

Material	Swartzentruber	Andy Weaver	Old Order	New Order	English
Plastic	30.0	2.4	20.3	21.6	81.7
Aluminum	40.0	43.9	49.3	37.8	80.8
Metal	80.0	26.8	42.7	37.8	70.4
Glass	40.0	0.0	18.7	8.1	74.6
Cardboard	10.0	9.8	10.7	8.1	73.2
Plastic bags	0.0	4.9	14.7	8.1	69.0
Paper	10.0	9.8	28.0	27.0	83.6
Styrofoam	0.0	0.0	0.0	2.7	22.1
Small batteries	10.0	0.0	4.0	0.0	36.6
12-volt batteries	30.0	53.7	61.3	54.1	30.0
n	10	41	76	37	213

is not as serious about recycling as some of our non-Amish neighbors, but on the other hand we have a non-Amish neighbor who throws it all in the trash and thinks recycling is for hippies."

There are some structural reasons why it's hard for Amish to participate in conventional recycling programs. Areas where Amish live are less likely to have robust recycling programs for household, business, or farm waste. Amish produce farmers use large bales of plastic mulch, but unless an agricultural-plastics recycling program exists, disposal by incineration or sending it to the landfill is the norm. It's also very inconvenient to take recycling materials by horse and buggy to local drop-off stations, especially in the winter. A relative lack of trust in the integrity and efficiency of the recycling system itself also plays a role. One shop owner who receives product in cardboard boxes recounted at great lengths his futile attempts to find a viable way to recycle the mounds of cardboard, finally deciding that sending it to the landfill was the least environmentally harmful option. Most Amish families have a burn pile, where paper, cardboard, and sometimes plastic are incinerated. For objects that are hard to burn, a trash pile on the back acreage is not unusual.

Although environmentalism and recycling are relatively new behaviors associated with the external world, the Amish community is not immune to the trends toward "green" living that are apparent in the broader society. The younger generations "don't create as much waste as we do," said one Old Order woodworker. "We used to have a burn pile and burned everything. Now people live a lot closer to each other, and trash blowing off the pile is a mess." Several Amish we spoke with also indicated that they realized that their neighbors disapproved of their burning plastic. Perhaps because of its visibility, many Amish look askance at littering. Recounting a trip to

Florida, one woman wrote to the *Budget* that she had noticed a lot of litter along the roadway. "Four of the Scenic Knoll School fathers and the boys took an afternoon off to gather trash along the road. They gathered 25 big garbage bags full and didn't get very far. One has to wonder why all the littering. Why not take it home and take care of it?"[46] But other "green practices" have had little success in the Amish community. One former Amish woman expressed her frustration with her parents and siblings, who mock her for carrying reusable canvas bags to the grocery store. "So then I have to pull the Creation Care thing on them," she laughed, referring to the faith-based environmental activism that has arisen in some evangelical and mainstream Christian circles.

Environmentalism and the Antiregulatory Amish

The case studies examined above illustrate the difficulty of categorizing Amish responses to local and regional environmental issues. It is true that Amish discourses around environmentalism exhibit a deep suspicion about the environmentalist agenda on several grounds. Their immediate response is to associate environmentalists with activists and extremists. Even Amish individuals who express sympathy for the need to protect natural areas almost always qualify their statements by distancing themselves from the "tree huggers" and the "animal rights people." The confrontational tactics employed by groups like Greenpeace or PETA go against Amish preferences for a more cooperative approach, but the Amish can also be critical of the messaging from more mainstream environmental organizations. "What I see from the Sierra Club is alarmist," stated a self-described Amish environmentalist in Montana. One result of this association of environmentalism with extremism is that many Amish downplay the seriousness of regional environmental problems relative to other issues. They feel that environmental groups have blown things out of proportion.

In addition, many Amish associate environmentalism with liberalism and governmental overreach. A man from the Andy Weaver affiliation was even more blunt: "For us, 'liberal' somehow gives a feeling of immorality, so if it's environmental or liberal, we don't want anything to do with it." That environmentalism equates with liberal extremism is based on the fundamental Amish assumption that not all forms of life have the same standing in God's eyes. "A Christian wants to preserve the environment which is God's creation," noted an Old Order minister. "But they cross over if they are more interested in preserving a frog over a human." Many Amish see the Environmental Protection Agency as epitomizing governmental overreach that privileges ecological concerns over human needs. "Well, the EPA has a bad name, of course, especially in our community," said one schoolteacher. "You hear that, and it's automatically negative." In Amish discourse, environmentalism is usually equated with locking up resources and forcing landowners to spend money on changes or behaviors the Amish think are unnecessary. The overriding perception of environmentalists is that they impede the Amish way of life.

Even as many Amish distance themselves from the discourse of environmentalism, they hold to an alternative narrative in which they are the ones who are "back to nature" and who use "common sense" to be effective stewards of the environment. "Environment is a big concern [for us]," said one man, and then in the same breath he noted that "environmentalists are on the back page." He continued, "I feel bad cutting down a tree, but I'm not a tree-hugger. I'd do it to pay my bills. But I'd be selective, to protect habitat for a cerulean warbler." In the environmental imagination of the Amish, stewardship means using resources like shale gas and forests responsibly, maintaining rural land and livelihoods in the face of urban development, and ensuring that their families are safe and well supported. When such choices are underwritten by Christian values and invested with their understanding of common sense, the Amish version of stewardship looks strikingly similar to religiously rooted, early-twentieth-century conservationism, which saw itself as "fundamentally linked with culturally inculcated ideas about human flourishing."[47] Reflecting this biblically inspired ethic of stewardship, one Pennsylvania man commented, "It would be a real shame if we didn't leave the earth in as good a shape as we found it."

To many contemporary environmentalists, the Amish narrative seems woefully inadequate for dealing with the array of ecological problems that plague the regions where the Amish live. As we show in chapter 11, the Amish are significantly less likely than most other groups to report proenvironmental attitudes. A 2017 study by Jessica Ulrich-Schad, Caroline Brock, and Linda Prokopy found that Amish farmers in northern Indiana were less likely than non-Amish farmers to be aware of water-impairment issues in and beyond their communities. Amish farmers were also less likely to adopt BMPs, including conservation tillage, grass or tree buffers, and nutrient-management plans.[48]

Amish skepticism of environmentalist agendas, however, does not automatically preclude Amish involvement in solving local and regional environmental problems. To the contrary, the watershed-remediation programs show that change can occur when non-Amish who have legitimacy in the community work together with Amish opinion leaders to soft-sell reasonable environmental initiatives that have clear local benefits. Reflecting on what he perceived as the heavy-handedness of the EPA, a Nebraska Amish man from Big Valley, Pennsylvania, confessed, "You're scared of 'em, you know? I think they should be a little more user-friendly." He suggested this approach: "Come in and teach us. You can catch a lot more flies with honey." Far from adopting the logic of environmentalism as their own, then, the Amish have maintained a critical view of the methods and fruits of scientific conservation. Their anthropocentric, religiously inspired ethic of stewardship is largely oriented toward community needs. But with the right approach and economic incentives, many Amish are willing to collaborate to solve environmental problems as part of their larger exchange with the outside world, which also values, though in different ways, the landscapes they know and inhabit.

Thinking Globally

The Amish Ecological Imagination

"Global" is one of the last adjectives one might expect to be used to modify the noun "Amish." Amish values are quintessentially local: love of family, commitment to church community, embrace of separateness, simplicity, and self-sufficiency. Even their history, as a "people apart," has reinforced their inward-looking nature. It seems somehow unfair to measure Amish attitudes against the challenges of global environmental issues, which are apparently so distant from their points of reference. But small though they are, Amish communities are indisputably part of a larger ecosystem, and there is a continuous connection between the local and the global. Small changes, multiplied by many actors, can have large-scale impacts. Amish relationships with the global environment therefore merit our consideration. Issues that compel some Amish engagement at the *local* level, such as pollution, are also issues at the *global* level. Of more concern are those human-generated problems that environmentalists argue are affecting not just communities but the entire world: loss of biodiversity, climate change, and population pressures on global resources.[1]

Appreciating the scope and complexity of global environmental challenges can be daunting for anyone, especially when, as for the Amish, they are outside their normal frame of reference. A planetary perspective requires access to information and a willingness to educate oneself about the realities that information describes. One must understand, or at least trust, information from scientists about how atmosphere, ocean, and land are interconnected, how biological diversity is distributed, and how human actions impact the physical and biological world. Large-scale thinking requires generalizing from what one can see to what one must assume by logical extension to be true. As humans, we are most comfortable at the level of direct experience. Even cosmopolitan Americans who have traveled widely and are attentive to international news, scientific advances, and cultural diversity can be overwhelmed by the scope of the global environmental challenges. The Amish lack many of those avenues of experience, but changes are drawing them into engagement with larger economic and ethical landscapes, both central elements of global environmental issues.

In this chapter we ask how the Amish understand their connections to nature and the human community at the planetary scale. We describe conversations with Amish from across the spectrum about four global environmental issues: land use and protection, preservation of biological diversity, evidence for a changing climate, and problems of a growing human population. We then compare responses from Amish and English residents in the Holmes County settlement to responses from a survey that has been widely used to measure environmentally friendly attitudes. We explore areas of agreement and difference and suggest ways that Amish history and theology seem to shape Amish perspectives on global environmental issues. Finally, we reflect on the challenges that make it difficult for the Amish to recognize and respond to global environmental concerns.

Land Use

In a world with an exponentially growing human population, the tension between development and land preservation is central to global environmental problems. When people appropriate land for farming, grazing, mining, and urbanization, natural habitats disappear. This creates conflict between the needs of humans and those of nonhuman species. Amish populations are growing in precisely those parts of the country where development pressures on the land are also strong. We wondered how Amish landowners would respond to local and regional programs to set aside land for the public good and how that might translate into support for land protection at the global level.

For the Amish, land ownership confers social status, and land is considered a wise investment. Historically, as Elizabeth Place reminds us, "rural existence has permitted the Amish to use their land freely with little interference from the state."[2] If there is a need for restraints on rural development, one local mechanism is zoning. But even in the early 1990s, John Hostetler noted that many Amish view zoning as a mixed blessing.[3] For example, if Amish want to operate small businesses from their properties, their plans may conflict with regulations limiting the number or size of outbuildings on a certain acreage. Some Amish shops generate noise and traffic from delivery trucks, and non-Amish neighbors may object. The Amish have been known to push back when zoning laws restricted their plans to expand their businesses, resisting governmental attempts to regulate land use even when it served the wider public interest.[4]

Within their community, Amish landowners value certain kinds of habitat diversity. They appreciate forests, where they can hunt, tap maple syrup, harvest firewood and mushrooms, and watch birds. Streams and lakes provide fishing and

The conversion of an old railroad track into a buggy and bike trail in the Holmes County settlement was accomplished with support from many in the local Amish community. Photo by Doyle Yoder

recreation. Some Amish have purchased hunting land distant from their homes, where they may have a small cabin or family campsite and might harvest trees for timber income. But the value of varied landscapes comes largely from the multiple ways they can be used. Although Amish birders value wetlands for shorebird sightings, other Amish generally regard swamps and marshes as breeding grounds for snakes and mosquitoes. With rare exceptions, the Amish do not own land that cannot be utilized in tangible ways.

They do, however, make active use of existing public lands. A Lancaster Amish man praised the extensive Pennsylvania game lands in his area because they provide access both to natural beauty and to hunting areas. Rails to Trails initiatives, converting old railroad lines into public recreational paths, have been organized in Amish enclaves in Ohio and Indiana, and they show the importance of local engagement in protecting land for public use. In Ohio, an Amish bishop was on the board of the Holmes County Rails to Trails Coalition, which paved the way, literally and figuratively, for a fifteen-mile trail from Fredericksburg to Killbuck. The project included an asphalt lane for bicycles and pedestrians and a chip-and-seal lane for horses and buggies. A board member told us that "most of the opposition came from non-Amish" concerned about losing their privacy and about noise and vandalism from trail users. But because the proposal included a lane for horses and buggies and, not coincidentally, ended its first segment at the Walmart in Millersburg, the Amish community embraced it as a way to avoid the traffic on Route 83. The trail has become a recreational magnet for Amish and non-Amish alike. "Initially we wanted it to get to town," said an Old Order Amish man, "but we've come to love it for Sundays and it's where the girls got (buggy) driver's ed."

The nineteen-mile Pumpkinvine Trail, through the heart of the Elkhart-LaGrange settlement, is also successful, but it initially encountered stiff opposition mostly from Amish landowners. At a public meeting, concerns were raised about "vandalism, loss of privacy, livestock welfare," according to one non-Amish man in attendance. The steering committee, which had no Amish members, was unsympathetic. "We sued the Amish and everyone else to get the cloud off the titles," remembered one committee member. This heavy-handed approach frustrated the Amish community. "Some of the initiators . . . came across a little rough with the landowners," said one Amish observer. "If they'd worded it a little bit better, or didn't tell them, 'Now you're going to have to do this, this, and this,' you know, there would have been a little more support for it." Many landowners who initially were skeptical, however, have warmed to the plan. Amish-run bicycle shops and other businesses have even sprung up along the trail, and while it does not end at the Walmart, the Dairy Queen right off the trail in Middlebury did so much business after the trail was completed that it embarked on a major expansion.

Habitat and Biodiversity Protection

In addition to providing land for public recreation, a major reason for land preservation is to protect critical habitats for plant and animal species. Sometimes this means setting aside sizeable areas where human activities are strongly limited. But many citizens, Amish and non-Amish alike, tend to assume that all land preservation is designed primarily to fence humans out. In fact, most public lands are open to recreation, and many are also available for resource harvesting. There are substantial differences between national parks, state parks, wildlife refuges, national forests, and other public lands. Nearly three-quarters of federal lands are managed through the Forest Service and the Bureau of Land Management for multiple uses, including not only recreation but also hunting, grazing, timber harvesting, mining, and other forms of resource extraction. In contrast, the Fish and Wildlife Service and the National Park Service administer federal lands primarily for habitat protection, species preservation, and recreation.[5] These parks, wildlife refuges, and reserves are of concern to the Amish because they limit human access and are strictly regulated by government agencies. Protected landscapes thus sometimes precipitate competition between human desires and the needs of endangered species.

In North America, the most recognizable symbol of natural beauty and habitat protection is the National Park System. While there are relatively few national parks in the eastern United States, they are abundant—and dramatically scenic—in the West. These landscapes protect habitats and species, but they also allow human recreational use. The national parks are bucket-list destinations for many Amish travelers, who value the beauty and uniqueness of the Grand Canyon, Yosemite, and the Great Smoky Mountains. An Old Order man effused: "The wilderness areas are huge and gorgeous. It's necessary to preserve these big tracts of land, and as individuals we can't preserve land like that." It's much easier to embrace large protected areas when they are spectacularly beautiful and somewhere else. One Old Order man told us, "Don't mess up our Alaska. I've never seen it, but most Amish would agree." National parks are islands of iconic scenery, valuable as destinations for a natural experience.

In the Amish world-view, however, the defining value of land is that it supports human needs. Land that is "locked up" contradicts the Amish understanding of the value of nature. While parks are admired for their spectacular scenery, they still hold land out of the reach of human needs, provoking criticism. A New Order man noted, "We should use creation in such a way that future generations will also have access to these things. To take a position that we shouldn't cut any trees down in certain areas . . . that's not really viable, because it was given for the well-being, the use of mankind. So that qualifies our idea of stewardship." A similar, strong inclination toward human use underpinned almost every statement we heard about wilderness in our conversations with the Amish. One Old Order timberman mused, "It comes down to, why do we truly believe it's here? Why is that Grand Canyon here? Or the bug, or the deer, or the bison? Has it just happened? Or was man here, and God put it there for him? We firmly believe that all that we see was put here for mankind, and it's not the opposite. It's there for our use. And not abuse." For the Amish, human needs ultimately determine how we act toward nature. However, "need" is a subjective and negotiable term. The idea that only land unusable by humans should be protected conflicts with the idea that other species are inherently valuable and essential to functioning ecosystems.

Because 92 percent of federally owned lands lie in the western United States and Alaska,[6] Amish in Montana, Idaho, Wyoming, and Colorado are surrounded by public lands of various types. Regulations govern their use, and permits are required for camping, hunting, timber cutting, and other activities. But federal land managers actively promote appropriate use of public lands. Though the Amish in St. Ignatius, Montana, live within the Flathead Indian Reservation, which has strict rules limiting nontribal access to the land, nearby national forests and the Bob Marshall Wilderness are readily accessible. A woodworker in St. Ignatius said, "We have huge protected areas, and I support keeping those, 100 percent, because we also have a lot of forest service lands that are open to logging and hunting." Like many Americans, the Amish still complain about regulations, but they often agree that limits are needed to protect against overuse. An Amish resident of Rexford, Montana, told us, "I think that places like the Redwoods, Glacier, the Bob Marshall, they have a role. There's so many people that would go in and demolish big areas just to make money. If we wouldn't have some wilderness places, there wouldn't be a balance."

Like their eastern counterparts, however, western Amish still are frustrated when animals take precedence over humans. The San Luis Valley in Colorado is home to two Amish settlements and three national wildlife refuges. An Amish farmer and his wife living in the valley told us, "As farmers, one thing that rubs us the wrong way is our irrigation wells get shut off from November first until April first, but [the Fish and Wildlife Service] can pump any time they want, for the birds. So that's not quite using common sense. As far as wilderness areas, I think we need to have them. But having thousands of acres for, just birds, you know—it doesn't really do much for the birds. They land in the refuge, but then come scavenge in the grain fields." Allowing birds access to resources that humans are denied is seen as contrary to what nature is all about.

Amish from the densely populated eastern settlements who visit these western landscapes also see them through the lens of limits on human use. Describing his trip west, an Old Order sawmill owner said, "I visited national parks. I appreciate them. But not all are managed as a good steward. Sequoia, Redwoods—by preserving some, the future generations can appreciate them. But there are millions of acres [under protection] in the west, and if a tree falls, it rots there. I know it adds to the ecosystem, but it's wasteful. I don't believe in going in and raping a forest—you should protect—but selective harvesting allows better production." The Amish believe that this use-oriented view of nature is common sense and avoids extremes. One Old Order man put it this way: "There's a ditch on each side of the road. So if you stay in the center, you're probably ok." Another man agreed: "You can have extreme left-hand and right-handed views. I think a good solid Christian will walk right down the middle." These statements also reflect a tendency among the Amish to see environmentalists as extremists. The Amish may accept protection—to a point. But their firm conviction is that landscapes have value primarily when they are managed for human use.

Management for human use, however, is seldom the recommended course for protecting threatened plants and animals. Many declining species need large tracts of undisturbed land. Most Amish say that animals should be protected if possible, since they are part of God's creation. An Old Order birder agreed that animals that are not directly useful to people "should still be appreciated, and there should still be areas where those species can live." All animals may be "good" in the abstract, but the Amish show the same frustration as many non-Amish when protecting an animal, especially a small, insignificant, or unbeautiful one, interferes with human choices. An Old Order man commented: "I do see the value of preserving lands for other species. But the Holmes County Trail was stopped just below Glenmont because they found some big dead trees, and there are some exotic bats who roost in those trees. OK, that's a little foreign to us. I'm not really sure of what it is, but I don't have an appreciation for that. I mean, can't we just cut down the trees, and that would eliminate the problem?"[7] An Amish forester expressed similar frustration about protection of biological diversity that in his mind goes too far: "It's just that there's a limit. Bats aren't pretty. And what value does a bat have? I don't know. There's a food chain here, whether you like it or not, and we're at the top." While scientists point out that it is the "little things that run the world,"[8] most people are intrigued only by large, charismatic species with which they can readily identify. The idea that humans should curtail their use of a habitat because of a bat or a beetle seems to them absurd.

For the Amish and many non-Amish, the gulf between "environmental extremism" and "common sense" is best symbolized by an owl.[9] The controversy in Oregon and Washington between loggers and conservationists over the northern spotted owl occurred almost three decades ago, but owls were repeatedly mentioned by those we talked with as an example of "environmental insanity." Without being clear on details, virtually every Amish person knows that to protect an owl the government locked up the forests, and loggers lost their jobs. Said one New Order minister: "Most Amish men would be outraged if lumber prices were increasing to protect the spotted owl. When men worship the Creature more than the Creator, this is grounds for [church] censorship. If your environmental bent affects too many people negatively, it is wrong." This notion of worshipping the creation more than the Creator, summarizes where the Amish see the tipping point and calls attention to the powerful social pressures that constrain environmentally inclined Amish. When asked if plants and animals had as much right to exist as humans did, a New Order man turned the question around. "Well, what do you think? We privilege humans. Otherwise we should just open the doors and let in all the flies. The German [version of the Bible] says, 'to control nature.' So if humans have a soul and animals don't, then that changes everything." In this view, when human needs conflict with those of other species, humans come first.

An Old Order birder reflecting on changes in his rural neighborhood said pensively, "I have fewer places that I can go without running into houses. Some of my favorite places [to bird] are not as good anymore. But on the other hand, I might someday need to build a house for myself or one of my children, and I'll do the same thing. It's just the reality of things." Many thoughtful environmentalists would agree that valuing biodiversity need not imply that other species are always equal to humans. There will be situations in which humans should take precedence, but, they would argue, humans have a duty to other species. We should use resources carefully and look for alternatives that best protect the rest of the world's biodiversity. The Amish would generally take a stronger anthropocentric stance: there is no harm in protecting land as long as there is enough, but deferring land use or altering their way of life to keep an owl or a bat from extinction goes too far.

Climate Change

As farmers, gardeners, and foresters, Amish are believed to be close observers of nature, noticing when the purple martins return and the crocuses bloom. Who better to document the gradual but pervasive shifts in natural events over the past decades that have been called "fingerprints of climate change"?[10] Scientists have tallied a long list of phenomena that are changing as a result of higher mean temperatures caused by increases in atmospheric greenhouse-gas concentrations.[11] The Amish, however, are generally climate skeptics. "It's not just the people that laugh," an Old Order man told us, "it's Mother Nature." Most Amish agreed. An Old Order forester told us: "There's a lot of skepticism about global warming. Personally, I believe that climate can change, but I'm not that concerned about it. I don't think you have to worry about it for the next thousand years. That's how a lot of Amish feel about it. Or they totally disbelieve that there can be any change at all." The alarm in some quarters of the English community is seen as part of an environmental agenda that most Amish patently reject.

While the Amish offer a range of opinions on the reality of climate change, we found no one who was seriously concerned about it. Those we spoke with readily agreed that the weather changed from year to year, but they said it had always been that way. They saw no evidence of global warming. Amish in the higher affiliations were more aware of the arguments for human-caused climate change but still didn't take it seriously. They shared the skepticism of some non-Amish about the data reported in the media, and their own observations did not convince them of the reality of a warming world. When asked about it, a young schoolteacher said, "I don't see a trend." An Old Order homemaker told us, "There's been a lot of fuss about melting, but I question if it's as bad as they say." A New Order minister declared, "It's more political than environmental," while another New Order man questioned the data and gave a theological argument to support his position:

> Global warming? Well, I don't believe in that, generally. I think that's what you'd call a "hoax." I'm not saying there's not some warming going on, but we don't have records for nearly long enough. They're meaningless. The scientists either don't know nearly as much as they think they do, or they don't know. And we have the Bible. When Noah came out of the ark there were major changes in the environment. And God said "There will be heat, there will be cold. As long as the earth stands there will be seed time and harvest." That to me is much more sure than anything an environmentalist says.

A former Swartzentruber man was similarly unconcerned. "If there is such a thing as global warming," he said, "the Amish would say, it's in God's hands." These comments echo some of the same doubts and counterarguments voiced by non-Amish who deny evidence of global climate change.

Although they see no warming trends, many Amish do agree that storms have become more frequent and more powerful in recent years. This is one scientific prediction of a warming atmosphere,[12] but many Amish gave more theological explanations. A New Order bishop said, "It seems like we have more extremes in the weather—long winter, strong droughts, lots of rain. Some people say it's all the things we're doing in space, but I don't know. I think it's a sign of the times, the Lord saying who's in control." Another New Order minister echoed that idea. "Some would say that the Bible says that there will be hurricanes and tornados in diverse places," he noted, "and so unsettled weather is part of end-time prophecy." An Old Order woman had a more novel interpretation. "I'm not educated in this," she said. "But I've heard that the lightning and big thunderstorms would 'burn up' pollution, and I like that idea."

If the science is correct, climate change will alter the earth. That idea in itself

Most Amish are skeptical of the scientific consensus that humans are contributing to global climate change, and some explain the increase in storms and dramatic weather events as a sign of God's displeasure. Photo by Doyle Yoder

provokes skepticism among many Amish. Within a biblical framework, God made the heavens and the earth and controls the world. Humans are powerless to alter global processes. God's promise after the Flood was cited repeatedly as evidence that there was no crisis. A New Order mother of five told us, "Of course we want to be good stewards of the land and take care of the environment, but then we hear about people raising some kinds of concerns, and we think, well, that's ridiculous. God's in control. He knows what's going on, he can do anything." Many questioned the scientific evidence. One man said, "I heard one of the men in our church say that it is true, that the North Pole isn't as thick, and it's warmer, but what the scientists are not telling you is that the South Pole is colder. I don't know, is that true or not, but of course my radar accepts that pretty quickly because it fits my world view." Some Amish claim ignorance of the arguments or simply deflect the question. A man who recently left the Swartzentruber Amish asserted, "Conservative Amish are not only ignorant of global issues, but they don't care." An Old Order furniture maker summarized the Amish viewpoint: "We're uneducated in our understanding of global warming and things like that. We believe God created the earth to be like Psalms 104. The earth is possibly going to grow old, and man will probably have a role in that to an extent. But the rain forests, and stuff like that, are a little foreign to us. They are not things that we have interactions with. We would trust that God has a perfect design for the earth." If God's design is for warmer temperatures and stronger storms, actions on our part are neither useful nor necessary. The Amish are steadfast in their belief that God is in charge.

Human Population Growth

"It would be nice if we knew how many people there were on earth," commented a Swartzentruber Amish man when we broached the subject of population growth. To non-Amish, who are used to seeing global statistics reported in the media, this might be a surprising comment. In fact, a New Order minister made almost an identical observation in a later conversation. "In my world," he told us, "people would ask, 'How do they know how many people there are?'" These questions are informative because they suggest that even pondering global population is an exercise distant from the minds of most Amish. Furthermore, it's hard for Amish to imagine how such numbers might be obtained. The tools of national censuses, satellite data, and statistical inference are outside their experience. While Amish communities are well versed in the numbers that describe their own church districts, the size of the world population is something they seldom think about.

At its current growth rate of 1.09 percent, the global human population of 7.6 billion will double in sixty-four years.[13] The current doubling time for the Amish population is about 21.5 years, three times the world average.[14] Family size is declining among more progressive Amish (see table 3.1), but their growth rates are still higher than those of almost any other group in the United States. As a result, Amish settlements have more people, more homes, and less open space. Church districts divide. Land becomes increasingly expensive. "Holmes County is full up," said a New Order bishop. "As a young person, we looked for the best site for a house, but now any place is a house site, even over a ravine. The houses around, the sounds of lawnmowers—it's disruptive." He regrets this. "We're not bitter. But we'd have it otherwise." When asked if this has prompted any discussion about limiting family size, the bishop smiled. "No, no, that's not going to be discussed."

One response to this crowding is to relocate. New settlements arise regularly, as families move to less densely populated locations both east and west; one man described it as "moving to the edges." For the New Order Amish, new settlements are often seen as a missionary calling. A greenhouse owner from Montana observed,

"The New Order should be starting a new outreach every year. It would be good for the young people, give them a sense of purpose in life." "Outreach" is a term applied by New Order communities to offshoot settlements established under the oversight of an existing church district, partly to reduce crowding and partly to remove children from peer pressure from youths in churches with different Ordnungs. In other Amish affiliations, establishing a new community may be less scripted and based more on individual family choices. One major impetus is a search for affordable land. In established settlements, land prices are often so high that young families cannot afford to purchase enough land for an economically viable farm. In the period from 1990 to 2000, one new Amish settlement was established every five weeks on average. By the second decade of the twenty-first century the rate was one new settlement every 3.5 weeks.[15] Their rapid growth means that Amish communities are springing up in places where they have never been found before.[16]

Nature is an important factor in the strategic location of new settlements. Before moving, members of migrating households visit various sites to assess their suitability. They seek a rural area with fertile soil, a suitable growing season, temperatures good for gardening, and available forest resources. St. Ignatius, Montana, was chosen in part because it had a "banana-belt" climate despite its northern Rocky Mountain location.[17] The spatial location of the community is also carefully considered. Amish look for areas with enough available land in nearby parcels for multiple households. In the western states, Amish communities are often located several miles outside an existing town that offers shopping, a hardware store, veterinary services, and healthcare options; the distance provides separation but can be traversed in a buggy. It's a plus if the non-Amish residents of the area are welcoming to an influx of Amish families. In the eastern United States, where depopulation of rural landscapes has left many small towns aging and economically stressed, the entrepreneurship and energy of young Amish families are often embraced by local town councils and township commissions as a way of revitalizing their community. Thus, the spatial location of an Amish community reflects a savvy assessment of qualities that will enhance the economic success of the Amish and protect their cultural distinctiveness.

The proliferation of new Amish settlements in part reflects the fact that commitment to large families is at the core of their belief system. When asked if he thought there were too many people on earth, an Andy Weaver man with eight daughters laughed, "That goes way over my head. I know there's a lot of human beings, but that's just fine." The Amish also see themselves as a small, insignificant element of the American and global landscape. A New Order minister and father of nine said,

"We see having a family as Kingdom Work. We're building Christ's kingdom on this earth. And this is unarticulated, but there is a sense that, the more children you have, the better job you're doing. A woman, especially—that is her work, for the Lord, having a family." An Old Order man echoed this sentiment: "I don't think it's our job to manage population growth, because that's something that God made, and controls." The understanding is that if God wanted Amish to have smaller families, conception rates would reflect this.

Amish ordained leaders endorse large families and dismiss the possible impacts of Amish population growth on the global environment.[18] Artificial birth control is discouraged, although the strength of this proscription varies among affiliations and church districts.[19] Individual Amish couples, however, may look for ways to pace their family growth. Although sexuality is intensely private in the Amish world, Amish women have a dense network of female relatives and friends in the community who share experiences and provide support.[20] A New Order minister told us, "Most Amish either use no form of birth control, or a natural form of birth control—not all, but most. There are intelligent ways to stall having children, that don't involve medication, and abortions, and all that, and we personally teach our young people [those methods]."

One difficult situation that might prompt couples to use birth control is the relatively high prevalence of inherited conditions found in Amish settlements throughout North America.[21] Although Amish couples generally do not undergo genetic testing before marriage, their family trees sometimes suggest that they are at risk for having affected children. While special-needs children are deeply loved and cared for in Amish families, some metabolic disorders can have disabling and heartbreaking consequences. According to a New Order woman, "Keeping from becoming pregnant is a huge thing within our culture. Mostly natural methods, using common sense. It's no sin to have [only] three children if you know the rest will [have serious developmental disabilities]."

The Amish generally regard family size and childbearing as expressing God's will for their lives. Although at the personal level Amish parents realize that a large family is costly, the idea that population puts a serious stress on global resources is foreign to Amish thinking. Children are a gift from God. With little understanding of global-resource distribution or population trends, the Amish do not connect their own reproductive choices with environmental stresses on food, water, or land at the international level. But even if they did, their faith rests in God's promise to provide for them.

As these conversations make clear, the Amish are generally unconcerned about

Although the average family size within many Amish affiliations is slowly decreasing, Amish families are still large, running counter to environmentalist concerns about the demands the human population puts on the earth. Photo by Doyle Yoder

the predicted ecological consequences of species loss, global change, and population growth that are widely discussed by environmentalists and educated urbanites. Some Amish avoid questions about environmental topics, saying, "I don't know much about that." Others engage the discussion but doubt the evidence or the urgency of the problem. Still others are openly dismissive, seeing the hand-wringing of global environmentalists as simply silly. As a people who espouse the "middle road," the Amish find the alarmist tone of the discourse off-putting. An Old Order man summed up the Amish perspective succinctly: "Our people don't think we need to worry about things going on on the other side of the world. If we're not involved with it, our Creator will take care of it."

Comparing Amish and English Assumptions about Nature

Our conversations with the Amish give us some insight into assumptions they might hold about nature, but their answers don't get us very far in trying to characterize their perspective on the environment. While we assumed that there would be differences among Amish affiliations, we expected a wider diversity in attitudes among the larger, more heterogeneous English population. To make such a comparison, we needed to apply a uniform metric to both the Amish and the English. We used the New Ecological Paradigm survey (NEP), created to measure personal beliefs that would foster an environmental world-view.[22] This set of fifteen questions has been used in the United States and other countries and thus makes possible comparisons between populations. Respondents indicate their degree of agreement to statements that probe the assumptions underlying global environmental issues. In our study of Amish and non-Amish respondents in northeastern Ohio, a total of 363 households answered at least some of the fifteen questions. Responses were tallied on a five-point scale, where the choices were "strongly agree" (5), "moderately agree" (4), "undecided" (3), "moderately disagree" (2), and "strongly disagree" (1). Average scores above 3.0 indicate progressively higher agreement with a proenvironmental stance, while scores below 3.0 indicate increasing levels of disagreement with the idea that humans are negatively impacting the global environment.[23]

Amish and non-Amish respondents were statistically different in their overall level of environmental-mindedness, as column 3 (mean total NEP score) in table 11.1 shows. On average, English respondents showed a positive tendency (3.47, or mildly positive) toward an environmental world-view. To put this into perspective, the mean total NEP score for fourteen comparable populations in the United States was slightly higher, at 3.58.[24] Of those fourteen groups, only three had NEP indexes lower than those we found among our non-Amish population; the lowest value in the fourteen samples was 3.22. As we expected, the English in our rural Ohio population also showed more variation than our three Amish groups (higher standard deviations), indicating a wider range of both positive and negative responses among the English. In contrast, the overall NEP scores for our three Amish groups, all about 2.7, indicated mildly negative attitudes and were less internally variable. The Amish in our sample were more environmentally disinclined than were those in all but one of sixty-nine studies reported by L. J. Hawcroft and T. L. Milfont.[25] That one study, of thirty-nine Anglo and twenty-eight acculturated Latino blue-collar workers in the United States in 2000, recorded values of 2.75 and 2.57, respectively, for these sample groups.[26]

The fifteen survey questions cluster around five underlying assumptions of an environmental world-view.[27] Each of these five "concepts" is addressed in three different questions on the overall survey, as shown in table 11.2. For four of the five concept areas, English respondents had significantly more positive attitudes than did their Amish neighbors.[28] But answers to specific questions also suggest that the

TABLE 11.1.
Results of the New Ecological Paradigm survey

Population	*n*	Mean total NEP score (SD)	Concept 1: Is nature fragile? (SD)	Concept 2: Are we heading for catastrophe? (SD)	Concept 3: Are humans subject to natural laws? (SD)	Concept 4: Are we running out of resources? (SD)	Concept 5: Are humans equal to other species? (SD)
Andy Weaver	38	2.72[a] (0.544)	3.15[a] (0.863)	2.44[a] (0.897)	3.89[ab] (0.506)	1.77[a] (0.672)	2.54[a] (0.960)
Old Order	73	2.70[a] (0.423)	3.10[a] (0.819)	2.44[a] (0.761)	3.79[a] (0.744)	1.76[a] (0.551)	2.50[a] (0.722)
New Order	37	2.76[a] (0.447)	3.21[a] (0.792)	2.43[a] (0.823)	4.19[b] (0.537)	1.71[a] (0.493)	2.32[a] (0.752)
English (Ohio)	215	3.47[b] (0.700)	3.84[b] (0.772)	3.52[b] (1.054)	3.74[a] (0.681)	2.98[b] (0.976)	3.32[b] (1.035)

Note: The four study populations were all from the same region of northeastern Ohio. Swartzentruber Amish families did not participate in this part of the survey. Means within each column sharing the same superscript are not significantly different from one another. Different superscripts within a column indicate statistically significant differences ($p < 0.05$). The concept groups are given in table 11.2.

Amish may have read some questions differently than did their English neighbors.[29] Asked whether human activities are leading toward an ecological catastrophe such as climate change, for instance, the English were significantly more persuaded that humans are having negative effects on earth's ecosystems. All three groups of Amish, in contrast, felt moderately strongly that an ecological crisis was exaggerated. The responses of the Amish might reflect their lack of confidence in science, or they might come from a conviction that the "liberal environmental agenda" is out of control. They might also reflect the Amish belief that God is in charge of the earth and human actions are too feeble to have global consequences.

With respect to the idea that humans are overusing earth's resources, the Amish disagreed that the human population was reaching a limit and felt very strongly that earth had plenty of resources. Although English scores were significantly higher than Amish scores, they were not unilaterally positive. The English mildly agreed that the population is reaching its limits, but they were undecided about resource exploitation. English answers suggest confidence that new technologies can solve our resource problems. The Amish, in contrast, would be more likely to express confidence that God will provide for his people as long as the earth lasts.

The assertion that humans should respect the needs of other species is central to global biodiversity protection. If all species are inherently valuable, we should protect and attempt to restore endangered species. The Amish mildly disagreed that other species should be considered in human decisions, while the English

TABLE 11.2.
The New Ecological Paradigm survey questions

Concept 1. Balance of Nature, or Is Nature Fragile?
When humans interfere with nature it often produces disastrous consequences (3)
The balance of nature is strong enough to cope with the impacts of modern industrial nations (8)
The balance of nature is very delicate and easily upset (13)

Concept 2. Ecocrisis, or Are We Heading for Catastrophe?
Humans are severely abusing the environment (5)
The so-called "ecological crisis" facing humankind has been greatly exaggerated (10)
If things were to continue on their present course, we will soon experience a major ecological catastrophe (15)

Concept 3. Antiexemptionalism, or Are Humans Subject to Natural Laws?
Human ingenuity will insure that we do NOT make the earth unlivable (4)
Despite our special abilities humans are still subject to the laws of nature (9)
Humans will eventually learn enough about how nature works to be able to control it (14)

Concept 4. Limits to Growth, or Are We Running Out of Resources?
We are approaching the limit of the number of people the earth can support (1)
The earth has plenty of natural resources if we just learn how to develop them (6)
The earth is like a spaceship with very limited room and resources (11)

Concept 5. Antianthropocentrism, or Are Humans Equal to Other Species?
Humans have the right to modify the natural environment to suit their needs (2)
Plants and animals have as much right as humans to exist (7)
Humans were meant to rule over the rest of nature (12)

Note: The number in parentheses following each statement indicates the statement's location within the survey instrument. Environmentalists would be expected to agree with statements shown in roman font and to disagree with statements in italics. Italicized statements were reverse-coded in scoring.

mildly (but significantly) supported the rights of nonhuman species.[30] The Amish also moderately to strongly agreed that humans were meant to rule over nature, although in conversations most Amish focused on stewardship over dominion. Overall, responses to this concept were consistent with opinions we heard from Amish regarding biodiversity and wilderness protection.

Both Amish and English respondents felt that nature was fragile and that human interventions into nature would be disastrous, but English respondents saw this as a significantly larger problem. The Amish believe that God's plan for the earth is paramount and humans should not interfere with that plan. Thus, to the extent that the Amish answers embraced the fragility of the balance of nature, it was not because humans can do whatever they choose but because in a biblical world humans should obey God's laws.

The assertion that humans are subject to the laws of nature produced agreement between Amish and English respondents, but for what we think are different reasons. In fact, the three Amish populations returned higher scores than did our English sample. All our respondents strongly agreed that human beings are ruled by the same laws as the rest of nature. Both groups strongly disagreed that humans might be able to control nature. The Amish responses for this concept were especially strong, perhaps because the laws of nature are God's laws and usurping them would be an act of hubris. From a secular, scientific perspective, these laws are equally as binding, but they are naturalistic and scientific. Because humans are a part of nature, the laws apply to people as well as plants and porpoises. In either case, both Amish and English respondents express views that would support an environmental outcome, despite what might be divergent reasons for their answers.

The results of the NEP survey highlight several points about the environmental attitudes of rural Ohioans, both Amish and English. First, Amish attitudes and assumptions about the world lead them to be less concerned about human impacts on the global environment. While the English in our survey were, by comparison, mildly environmentally conscious, they also had a low mean NEP compared with most other populations in the United States and across the globe.[31] Rural Ohioans, by our measure, do not display strong proenvironmental attitudes. However, the Amish and the English agree on two important points. First, humans are subject to the laws of nature and limited in their ability to bend nature to their desires. Second, humans are capable of making bad decisions that can have negative effects on the natural world. For both the Amish and the English, nature is organized and predictable. Whether this is so because it was created by God or because it is inherently beautiful and complex, this shared appreciation of nature's value may give all of us common ground on which to work for its continued integrity.

Global Environmental Attitudes

Embracing a global environmental world-view requires that people think of themselves as members of a community of living organisms surrounded by a finite natural world in which human actions will have consequences for our shared future. This assumption is embedded in the survey questions that measure "environmental concern." But for the most part, this is not the world-view of the Amish, for whom nature was created by and is governed by God for human use. Natural cycles, in their view, are largely independent of human actions. God manages global processes, and so asking how human behavior affects the global environment is not a useful exercise. Thus, it is not surprising that the Amish mostly dismiss the arguments and exhortations of environmentalists. Without a sense of agency about their role in shaping the earth's future, they have little reason to look beyond their own families and communities.

The Amish also believe that the earth is temporary and that the end could be near. While the strength of this conviction varies among church districts and families, it is an underlying element of Amish thought that tends to make taking a long view seem unnecessary. Speaking about the earth's resources, one Old Order man said, "A person's belief of what reality is, and what this earth is here for, will greatly dictate how we use it. Is it a permanent thing, or a temporary thing? Although we still want to take care of it, it's not going to be quite as much of a priority if it's just temporary." Amish limits on formal education, especially their distrust of parts of the scientific world-view, may also deter them from working through the evidence offered by scientists and policymakers. Their distance from the digital world and from many mainstream sources of information limits their exposure to such issues. Like many non-Amish, they are more persuaded by the opinions of their personal authority groups than by distant, declarative voices.

While those citizens who do embrace an environmental concern for global issues of climate, biodiversity, and population understand the importance of behaving in ways that are more sustainable, only a very few actually live a lifestyle that is truly "light on the earth." Environmental attention is hard to maintain over the long term. It requires constant self-examination, a degree of self-denial, and a continuous intellectual search that is wearying and sometimes seems futile. We are enmeshed in an economic, social, and political system that seems to diminish our individual efforts, so that it is easy to make choices that under closer examination are not environmentally benign. Even among those with a strong, grounded personal belief in a world-view that would make this effort worthwhile, most stray into the path of least resistance, following the deep, culturally carved ruts that mark the road, without asking where the road ultimately leads.

All of us, Amish and non-Amish alike, are most likely to be galvanized in our choices when we can plainly see that our decisions and actions make a difference. Those situations typically arise, not at the global scale, but within our communities or our regions, where we can share a common sense of the problem, where we can clearly see our neighbors who will be affected, and where we can collaborate on possible solutions. As we have seen, there have been situations in which the Amish were persuaded that they should alter or modify their traditional practices to address local environmental issues. Perhaps it is at this scale that the Amish are most likely to cultivate an environmental perspective, to build the bridges to their larger communities that would allow them to act locally and perhaps even think globally.

CHAPTER TWELVE

Parochial Stewards

The Amish Encounter with Nature and the Environment

The pastoral image of Amish communities still living simply and in touch with the land strikes a deep chord with many Americans. This nostalgic affection has many elements, but one may be that the Amish offer hope that humans can, in fact, live well, in relative harmony with their environment, without all the technologies and consumer goods that saturate life in industrialized societies. The Amish are not alone, of course, in being held up as subjects of ecological admiration. For similar reasons, idealized images of indigenous groups, such as Native Americans, with longstanding and deep ties to the land and a commitment to defending earth's resources remain powerful for the general public. Both groups are held up as models of sustainability even though for the Amish, as a white, Christian, agricultural people, the realities of engagement with nature diverge in important ways from those of indigenous societies, many of which were colonized by Western powers.[1]

But the Amish share another reality with many indigenous peoples. They all find themselves in new and rapidly changing political and economic circumstances, where their interests and their views of nature may diverge from the expectations of many outsiders, including environmentalists. Corporations run by Alaskan natives have entered into joint ventures with oil and mining companies, while Amazonian Indians have sold or leased portions of their land for timber and other natural resources.[2] When historically marginalized groups use their newfound autonomy and political muscle to pursue environmentally harmful economic development in ways that mirror the behavior of their colonizers, they are often accused of selling out their heritage. In a group that has been placed on a pedestal, normal human qualities of personal and collective self-interest come to be seen as a failure of global responsibility. In the case of the Amish, their gradual emergence as objects of positive public regard has come over the same half century during which many of them have left farming for new, nature-based extractive businesses and experienced unprecedented affluence, diversity, and mobility.

These new realities of Amish life greatly complicate the simplistic image of a people in tune with nature. In this changing context, their use-oriented approach to nature and their provincial outlook have sometimes led to friction with non-Amish environmental agencies. Without a doubt, the biblical admonition to "be not conformed to the world" has been instrumental in keeping Amish lifeways intact in the face of ever-encroaching outside pressures. But the goal of separation from the world is not to live lightly on an increasingly crowded and ecologically stressed earth. Its primary focus is on the well-being of Amish church districts and communities. It is therefore at odds with one of the most fundamental principles of ecology, namely, that everything is connected.[3] What happens locally has the potential to resonate in regional and global environments. In many ways, then, the Amish are parochial stewards of nature. In their emphasis on localness they embody a place-based closeness to nature that many environmentalists affirm, and their comparatively low rates of consumption mean that they demand less of the earth's resources. But the Amish diverge from ecological ideals in that their avowedly inward focus typically leads them to dismiss the idea that human actions can have a cumulatively negative impact on the global ecosystem.

In this concluding chapter, we try to make sense of the environmental implications of Amish world-views and the adaptations the Amish have made to a changing national and global context. We reflect on two larger issues raised by our case study: whether religion-infused, small-scale societies promote behaviors that are more respectful of the natural environment and whether such societies offer insights into the relation between growth-oriented capitalism and a sustainable ecosystem. And we ask why idealized understandings of the Amish as living in harmony with nature persist and reflect on how Amish and non-Amish might learn from each other and find common ground in the pursuit of wise stewardship of earth's resources.

Recapping Amish Interactions with Nature

Our study has shown, first and foremost, that diverse ways of engaging with nature exist across the Amish spectrum. Amish who have adopted chemical-intensive agriculture live side by side with those committed to organic farming. Households with minimal energy and consumption budgets coexist with families whose consumption of natural resources is on par with that of their rural, non-Amish neighbors. Some Amish spend their entire lives in the familiar landscapes where they grew up, while others travel widely to sites of natural interest or relocate to places all across the country. Some children come to see cats as expendable, while others grow up lavishing love on their pets. Amish who turn to natural remedies to treat a variety of minor and serious ailments coexist with those who elect to seek treatment in clinics and hospitals that practice science-based medicine. Dog and deer breeders defend their businesses as an appealing way to make a living on a small plot of land, even as other Amish worry that they push the limits of acceptable livelihoods. In countless ways, the Amish engagement with nature is shaped by affiliation, income, gender, occupation, history, and personal choice.

For all this diversity, however, the Amish are by and large united in seeing nature as existing for human use. Since they believe that only humans have souls, they generally see the protection of other living organisms and their habitat as secondary or inconsequential. Even asking how the Amish interact with nature, as we have done in this project, assumes that nature has an independent existence and that humans are relative newcomers to the natural landscape. In the Amish view, however, nature and humans were created only days apart. God gave people the garden and the earth, and the Amish find it almost impossible to think about nature as separate from humans. They generally believe that nature will endure as long as humans endure and such cosmic questions are best left in God's hands anyway. This strong anthropocentric notion that nature was created for human use infuses every aspect of the way the Amish think and talk about nature. In agriculture and in forestry, for instance, land that is not actively managed is seen as wasted or unproductive, especially if it is protected from human use. The religious underpinnings of the Amish world-view thus stand in stark contrast to an ecocentric philosophy that stresses the inherent value and mutual interrelatedness of all organisms and ecosystems.

In some respects, this anthropocentric and utilitarian view of nature is similar to that found in certain other white, religiously conservative rural communities.[4] Working in occupations that rely on natural resources may engender a nature-exploitative view of the world. What is distinctive in the Amish case is the intentionality of their selective withdrawal from institutions and discourses that are central to public life. This intentional separation leads to relatively low levels of interest in what's going on outside the boundaries of their communities and limited access to accurate information about environmental problems around the globe. Even as children, Amish absorb this cultural world-view from their home experiences. When asked to portray their neighborhood in a drawing, Amish children drew their own homes and family compounds but did not depict any neighbors' properties in their sketches.[5] While the doctrine of separation from the world leads the Amish to nurture a place-based sense of community that is closely tied to the land and to its resources, their local and parochial focus sometimes leads them to believe that what they have always done will always be fine. It thus creates the conditions for a stewardship that is admirable in some respects and yet limited and constrained in others. Although the Amish do "care about nature," they express this sense of stewardship contextually and often in competition with other interests.[6]

Given the theological underpinnings of their attitudes toward nature, it is easy to see why the Amish are often critical of post-1960s secular environmentalism, which Evan Berry argues "wanted to build practical, policy-driven solutions to environmental problems without getting caught up in the messiness of religious ethics."[7] Their understandings of nature, like those of the farmers in Nora Haenn's study of Calakmul, Mexico, make them critical of scientific attempts to measure and manage natural processes.[8] Moreover, the social distance the Amish perceive between themselves and politically liberal urban elites who are the core of the environmental movement increases their tendency to dismiss science-based reports about environmental challenges facing the earth and to ignore calls for personal and collective action to redress environmental damage.[9] When viewed from the perspective of environmentalism, Amish ways of thinking about and relating to nature can sometimes seem to be part of the problem, not part of the solution.

In many respects, early-twentieth-century conservation offers a more appropriate framework for understanding Amish views of nature. The "conservation of natural resources," a term popularized by Gifford Pinchot during the Progressive Era (1890–1920), was a program of land protection as well as a utilitarian enterprise that saw hunting, fishing, and extraction of natural resources as part and parcel of the wise use of public lands. And though conservationists saw science as crucial to their efforts, they championed an applied science that was blended with a concern for "the greatest good for the greatest number for the longest time."[10] The social values that surrounded conservationism were very much in line with Amish thinking, especially the gendered division of labor that rendered "productive" men as shapers of the fields and of the community environment and "nurturing" women as keepers of nature close to home.[11] Most importantly, the discourse of the conservation movement was saturated with religious overtones. Pinchot, himself of Protestant background, tied rurality, religion, and conservation to national development, which would not reach its potential "unless life in the country is vigorous and sound . . . and unless the country church does its full share to make it so."[12]

Such parallels must be drawn with caution, since the Amish are not stuck in the past. And the national and global context has greatly changed since the early 1900s. In the early years of the twentieth century, Pinchot supported President Theodore Roosevelt's creation of new public lands and was a vocal advocate of careful management. But even then, he realized that natural resources were finite, warning that it was "stupidly false" to consider the nation's resources inexhaustible.[13] It was only in the post-1930s era that global population began to escalate and new technologies were developed that could dramatically alter the natural environment. In spite of these new realities of global environmental degradation over the past century, we encountered many Amish who had mixed feelings about setting aside large wilderness reserves and who believed that God would continue to provide enough land and resources for everyone. Standing on his back deck, where he could get better reception for downloading the latest bestsellers on his Kindle, one Old Order man told us: "I consider myself a conservationist, not a preservationist."

The Amish and Religious Environmentalism

Ever since Lynn White Jr. asserted that Judeo-Christian cosmology carries the lion's share of blame for environmental problems facing the world, there has been a growing tendency among environmentalists and scholars alike to search for direct, causal links between religious doctrine and sound ecological practices.[14] In contrast to Christianity, Eastern and South Asian cosmologies (especially Daoism, Hinduism, and Buddhism) and indigenous traditions (including those of American Indians) have been held up as ecocentric because of "their assumptions of interconnectedness between human beings and nature."[15] Indigenous peoples in particular have been idealized by environmentalists as offering models of human society that use nature without compromising ecological integrity. Traditional ecological knowledge, or TEK, has become a popular field of study both as a way for outsiders to tap into indigenous wisdom and as a way for indigenous peoples to reclaim and assert their heritage.[16] According to Ben Orlove and Stephen Brush, indigenous peoples are frequently depicted as living in harmony with nature because of their long histories of living in ecosystems without seeming to disrupt them, a deep knowledge of nature and how to manage it, and "religious beliefs about ritual uses of animals and plants that safeguard their sustainable use."[17]

Yet indigenous cosmologies that stress interconnectedness may be far less environmentally friendly than they appear.[18] The historical reality of ecologically sustainable Asian religions or ecologically noble indigenous cultures has been questioned by a number of ethnographic and historical studies. For example, massive deforestation occurred in Japan twelve hundred years ago in spite of the adoption of Buddhism and the presence of an animistic native cosmology, Shinto. Some indigenous practices have clearly harmed the ecosystem, such as the harvesting of bird feathers for ceremonial use by the indigenous peoples of Hawaii, the poisoning of entire water holes in the Kalahari by G/wi foragers, or the overhunting of moas, giant swans, and other large, flightless birds by the Maori. One review of the literature concluded that "on balance, the evidence on faunal impacts of small-scale societies indicates that conservation is absent and depletion is sometimes a consequence."[19] None of these examples denies the much more massive ecological destruction wrought on small-scale societies by colonizing powers or discounts the traditional ecological knowledge that clearly flourished in such cultures. Rather, their sustainability was more a function of low population, limited technology, or other external factors than of an explicitly eco-friendly religious or philosophical orientation. Our case study of the Amish thus reminds us that it is important to distinguish between intention and outcome and that religion, culture, and nature are intertwined in complex ways.

Intention versus Outcome in Ecological Practices

At the individual level, the discrepancy between intent and action, known as the environmental values–behavior gap, has long been recognized. Participation in environmentally supportive behavior rarely mirrors the strength of one's stated commitment.[20] The Amish present a somewhat different case, which Eugene Hunn calls epiphenomenal conservation.[21] Behaviors that appear ecologically sensitive from the outside may be by-products of other factors rather than performed with environmental intent. Indeed, the restraints placed on technology and consumption by various Amish groups, which in turn produce behaviors that are less resource intensive, are not motivated by ecological intent. Instead, they are best conceived of as an ecological by-product of the religious doctrine of separation from the world. But relative to each other, how important are intention and outcome as components of an ecological world-view? Comparing the biblical literalism of the Amish with theology in other Christian denominations can help us answer this question. Christianity is strongly anthropocentric, but such an orientation to nature need not be an obstacle to ecological sensitivity.

Within the Catholic tradition, Pope Francis's 2015 encyclical, *Laudato si': On Care for Our Common Home*, begins with a clarion call to recognize that the environmental crisis is of our own making: "The earth, our home, is beginning to look more and more like an immense pile of filth."[22] In addition to Pope Francis's acceptance of the scientific consensus on climate change and his recognition that the poorest will suffer the most dire effects of environmental damage, the encyclical is noteworthy for its claims that species "have value in themselves."[23] The 2006 publication of *The Green Bible*, designed to highlight the bible's "powerful message for the earth," represents another high-profile, interdenominational effort to document a Christian ethic of sustainability.[24] Among evangelical groups, the Creation Care movement reframes the relation between God, humans, and nature to promote environmental protection. The Evangelical Environmental Network, for instance, helped defend the Endangered Species Act, characterizing it as "the Noah's Ark of our day."[25] Though diverse in their formulations, these new forms of Christian religious environmentalism all acknowledge that humans have created an environmental crisis and that science can help us understand and resolve it. And they promote a strong stewardship ethic based on the biblical assertion that God gave humans the earth and pronounced it good, independent of human needs, conferring value on nature in the most unambiguous way: by divine decree.[26]

These examples illustrate that an anthropocentric approach to nature is not necessarily at odds with environmentalism. One way to construct an effective environmental ethic on the basis of Western anthropocentrism would be for individuals, corporations, and other interest groups to "consider how their actions that directly affect the natural environment indirectly affect other human beings."[27] For example, using wood from tropical forests may benefit well-to-do consumers and the timber and furniture industry, but it may also "deprive indigenous peoples of their homes and traditional means of subsistence."[28] The Amish are well aware that inequalities resulting from resource extraction can undermine community life, as their concern that fracking leases could create wealth disparities and sow jealousy within their churches shows. But extending this ethic to non-Amish populations is a tall order. Concerning themselves with inequalities in the larger society runs counter to the very insularity that has helped the Amish survive. This inward focus is also reinforced by outward symbols of difference and by institutional structures such as adult baptism and excommunication, which lead community members to prioritize belonging over virtually everything else. Those Amish individuals who promote a stronger stewardship interpretation of Genesis and a broader definition of community to include non-Amish and nonhuman creatures still swim against powerful cultural currents.[29]

If intention is the primary criterion for evaluating ecological practice, then the views of the Amish hardly make them a model for environmentally conscious living. As we have seen, the Amish score very low on the New Ecological Paradigm scale, which has been used to measure environmental attitudes in different populations. But in the end it is behavior, more than attitude, that reduces human impact on the world. Many scientists and environmentalists have pointed out that talk is cheap when it comes to sustainability. In this respect, the somewhat lower carbon footprint of the Amish stands out as an important outcome even if it is not ecologically intended. Surrounded by a high-tech world, the Amish set conscious limits on resource use in many (but not all) aspects of their daily lives, some of which have important positive ecological results. In the final analysis, both intention and outcome are important aspects of ecological practice. Intention matters because over the long term it leads to attentiveness to environmental issues, provokes political action, and promotes research. Yet when intentions are divorced from behavior and actual outcomes, their power wanes considerably and opens those who talk the talk but don't walk the walk to charges of hypocrisy. Our study has shown that Amish culture and the Amish version of Christianity do not constitute an ecotheology, but they do exhibit many environmentally positive behaviors. The Amish show that a life well lived doesn't require endless consumption or owning the latest technology.

The Inseparability of Religion, Nature, and Culture

Religious cosmology in any society is closely intertwined with cultural and politico-economic forces. Amish skepticism toward environmentalism as a set of attitudes and policies is not just a function of the biblically inspired separation from the world. It is also fostered by Amish views of authority, Amish attitudes toward the state, and Amish cultural identity as rural farmers and entrepreneurs. Central among these is the importance of trust and relationships in translating the authority of the state and of the scientific establishment into something that is intelligible and acceptable within the Amish community. In cases where trust has been built across the Amish-English divide, we have seen some surprising success stories. Some Amish have been involved in watershed remediation projects or have participated in Rails to Trails initiatives. In such cases the establishment of meaningful, long-term relationships between Amish and non-Amish has been instrumental in laying the groundwork for change. More often than not, however, the Amish exhibit a low degree of trust in the basic institutions of government and science and their corresponding bureaucracies.

Amish skepticism of state authority arises both from their history of persecution and from their belief that large-scale bureaucracies are too impersonal, inefficient, and imbued with arbitrary power.[30] Some of these concerns are not unfounded. Though they usually have good intentions, governmental employees use a variety of bureaucratic techniques, including written reports and selective inattention, to exert their control in ways that may disproportionally affect minority populations.[31] Yet in spite of the state's uneven track record in safeguarding the public good, the modern environmental movement is in part based on the notion that the government should have a prominent watchdog and enforcement role. In addition, residents have a civic responsibility to be informed and to voice their opinions about how we should act toward nature. The Amish, however, are "ambivalent citizens," engaging with government more as "subjects" who believe that "each group of people should arrange its own distinct privileges and duties with the powers that be."[32] As we have seen, the Amish do come into contact with and are influenced by a variety of government or quasi-government workers: certified foresters, agricultural-extension agents, public-health commissioners, EPA officials. Because the Amish assume that government is linked to the exercise of force, however, trust between their community and government officials is tenuous and usually is built on a case-by-case basis.[33]

Amish attitudes toward science are similarly complex and often hinge on the perceived trustworthiness of the authority figures making a particular claim. While they sharply delimit the science curriculum in parochial schools and dispute the scientific consensus on the formation of earth and life on the planet, they accept certain scientific principles that are not at odds with biblical statements, especially when consensus develops among individuals in their primary reference group. Virtually all Amish accept that the earth is round or that gravity is a powerful force. Amish of every affiliation also judiciously use the fruits of science in medical treatments, animal husbandry, agriculture, shop technology, and much more. In this tendency to use science selectively, the Amish are no different from many non-Amish. They remind us that science is not a stand-alone, self-justifying authority; rather, the legitimacy of its findings is always granted culturally.

Science, however, is defined both by its fruits and by its methods; for scientists, the two are inseparable. The distinctiveness of the Amish approach lies in its rejection of the methodological assumptions underlying science, namely, that inquiry and hypothesis testing are a productive way of interacting with the world and that truth must be determined and governed by empirical observation, rather than relying on supernatural explanations. Many Christians do not see science and faith as inevitably in conflict, because they are comfortable with the possibility that God set the natural world in motion billions of years ago. The biblical literalism of the Amish, however, largely rules out this interpretation. For the Amish, the expulsion from Eden and the Flood mark the historical moments when humans most shaped the environmental conditions of the planet. As Nicole Welk-Joerger points out, these two falls are irreversible, life-altering moments in Christian doctrine, at least until the Second Coming.[34] So powerful is this narrative for the Amish that they almost always filter claims of secular science and environmentalism through it.

Rainbows have special significance for the Amish, for whom they symbolize God's covenant with Noah after the flood and imply that God will always provide for humankind. Photo by Doyle Yoder

The Amish, the Growth Economy, and Sustainability

The idea that the historical shift to industrialization and modernity in seventeenth-century Europe carried risks as well as opportunities was widely acknowledged by classical social theorists such as Karl Marx, Emile Durkheim, and Max Weber. Yet even they did not foresee that "the 'forces of production' would have large-scale destructive potential in relation to the material environment."[35] This unprecedented human refashioning of the earth's biosphere during the twentieth century has provoked much theorizing about the underlying causes.[36] One of the earliest formulas for evaluating the ecological consequences of modernity was the I=PAT equation. Formulated in the early 1970s as a more refined version of the Malthusian proposition that population growth would inevitably lead to food shortages, it proposed that environmental impact (I) arises from the interaction of three major factors: population (P), affluence (A), and technology (T).[37] In the decades since IPAT, numerous scholars have explored the relation between technological efficiencies, economic growth, and population as they bear on accelerated rates of species extinction, climate change, and pollution. It is beyond the scope of this study to provide a comprehensive analysis of these debates. We believe, however, that the Amish case is valuable for helping us think about the relation between growth and sustainability and how Amish and non-Amish might learn from each other in the pursuit of wise stewardship of earth's resources.

When Amish lifestyles are viewed against the I=PAT model, the Amish share more similarities with countries in the global South than with wealthier nations in the Northern Hemisphere. In terms of population, the average size of their families remains large and has declined only slightly as they have moved away from farming. And building God's kingdom on earth comes with significant ecological costs. According to a 2008 study by Oregon State University researchers, the summed carbon emissions of descendants far exceed the lifetime emissions produced by the original parent; thus, reproductive choices affect a person's ultimate impact on the environment.[38] In addition, over the past few decades the Amish as a whole have become more affluent and made more compromises with technology.[39] With the exception of the Swartzentruber Amish, whose pace of change is noticeably slower, most Amish are using, if not owning, more types of technology. And as our footprint study showed, they are becoming more consumer oriented. The outward signs of Amish-ness—the horses and buggies, the one-room schools, the plain clothing—are still intact. Underneath the hats and bonnets, however, lie some of the telling hallmarks of modernity, including leisure time, travel, and even social-class divisions. Tensions between ordained leaders and business owners who want to be free of church restrictions on technologies on weekdays are a growing dynamic in some church districts.[40] Our study reveals just how difficult it is, even for a self-proclaimed religious group that takes separation from the world as its guiding ethos, to maintain distance from the growth economy.

In spite of these far-reaching changes, however, the Amish remain distinctive in several ways that we believe can inform the wider conversation about how to make more sustainable choices for inhabiting the earth. For one, they demonstrate that self-provisioning is possible even in our twenty-first-century disposable society. The slogan "Use it up, wear it out, make do, or do without" is well known in Amish households. Their preparation and consumption of food, though undergoing changes, still illustrates many elements of a harvesting lifestyle. In spite of their increased reliance on Walmart and other retail outlets, they grow and preserve much of their food, and they often supplement what they grow and buy with fish, game, and other wild products. Because they are surrounded by fields, gardens, animals, and farms, the Amish know where their food comes from. Such a recognition is becoming increasingly important to a growing segment of the non-Amish population who look critically at the processed-food industry. The spread of farmers' markets, including those attended by Amish vendors, is a response to these consumers' demands for seasonal and locally grown foods.

The small scale of Amish institutions also stands out in a world where seemingly everything, from megafarms to megachurches, is scaled up to higher levels of efficiency and predictability. According to the sociologist George Ritzer, the resulting "McDonaldization of society" produces "a wide array of adverse effects on the environment."[41] While the Amish do not oppose economic growth, their shops, schools, and church districts are intentionally designed to keep interactions on a human scale. This is an important counterbalance to the concentration of economic and political power in ever-larger and inscrutable corporate and governmental entities. Large-scale organizations often detach decision making from production, and when they make bad decisions, the ecological consequences can be far-reaching. Moreover, large businesses invest vast amounts of money in advertising campaigns designed to encourage consumers to buy without thinking. In contrast, a vibrant network of small shops and businesses provides both an incentive to shop locally, keeping money in the community, and an ideal setting for apprenticeships for teenagers, who can work under the tutelage of caring adults. Economically struggling rural communities often find large, outside industries attractive, but as Wendell Berry reminds us, "The most effective means of local self-determination would be a well-developed local economy based upon the use and protection of local resources."[42]

Finally, the Amish offer a thought-provoking example of how systems of accountability can be set up to enforce community limits. As global-footprint analyses show, humans are clearly degrading the planet faster than we are regenerating it, yet finding the collective will to limit consumption and technology seems almost impossible. In spite of decades of pledges, protocols, and collective hand-wringing, the growth economy is well entrenched, fueled by large corporations and governments that are more and more difficult to hold accountable for their actions. The Amish restrictions, though not undertaken with ecological intent, represent a collective attempt to value family and community life over material things. They recognize that human wants often greatly exceed human needs, but then they go one step further. The secret to their ability to resist unbridled consumption and ownership of certain technologies lies in their community structure and in the idea of an Ordnung as a form of community accountability for behavior. In this sense, Amish practice echoes the most up-to-date consensus in the field of political ecology about decelerating the growth economy: focusing on individual intentions and voluntary simplicity is not enough given the magnitude of the planet's ecological distress.[43]

The irony, of course, is that the Ordnungs of Amish church districts are typically silent on explicitly environmental matters. But the concept of an environmental Ordnung is an intriguing one and worthy of wider debate. Individual choices, such as choosing paper over plastic or deciding to buy an electric car, are valuable. Institutional incentives such as tax breaks and market forces such as carbon offsets can also have a significant impact. But the lessons of the past few decades suggest that voluntary choices, institutional carrots, and market mechanisms will not be enough to turn the tide. Some form of laws or community pressures that have teeth in them seem to be needed to enforce ecological behaviors and outcomes. Within the Amish community, churches use social pressure up to and including excommunication and shunning, but even within their church districts these methods of social control are sometimes contested. The Amish do clearly show us the considerable power of a system of prescriptions and proscriptions agreed upon by all members of the community and backed up by authority figures who prioritize community interests. Even at the regional level, establishing a system of accountability and mechanisms for enforcement has been shown to make a difference in environmental outcomes.[44] In a diverse, secular world these institutional buy-ins might come from a variety of social, political, and religious entities. Given the scope and magnitude of the world's environmental problems, these elements of social persuasion will need to be developed at every level of government: from the United Nations Framework Convention on Climate Change to national, regional, and local initiatives.

Demystifying the Amish

In light of the far-reaching changes that have transformed Amish society over the past half century, why is it that we want to make the Amish into ecological role models when the reality is decidedly more mixed? Part of the answer, as David Weaver-Zercher points out, is that such stereotypes serve both ideological and pragmatic functions for many non-Amish audiences.[45] For some tourists, the Amish evoke a "pioneer" mentality that is powerful in its own right. The "old-fashioned" approach of the Amish is a reminder of how their grandparents and great-grandparents might have lived. In a world that sometimes seems to reel from one environmental crisis to another, the idea that people can live in relative self-sufficiency without modern conveniences is strangely reassuring. The possibility of "simplifying radically"—as Matthew and Nancy Sleeth did—is one that we can imagine and draw hope from. Yet relatively few non-Amish actually pursue a radically simpler lifestyle, suggesting that despite the nostalgia that pastoral stereotypes evoke, such images can also reinforce a sense of the superiority of modern life. At the end of the day, tourists can drive home in their air-conditioned vehicles content in the wisdom of their own choices.

For champions of small-scale agrarianism, images of the pastoral Amish can fulfill a slightly different function. By highlighting the loss in mainstream society of a close connection with the land, images of the Amish as subsistence farmers "serve as a powerful internal cultural critique."[46] The Amish example, thus caricatured, is deployed to exhort, cajole, and sometimes even shame the rest of us into rethinking our relationship to industrialization and its destructive consequences. Even though the Amish are changing, they remain among our last living examples of a society that consciously chooses to resist being co-opted by a hypermodern world. In this sense, the Amish represent a valuable element of human diversity to which we can turn for inspiration. We are hopeful that they can hold the line on technology and provide an example of how a fulfilling, community-oriented life can be based on living close to the land. The late agrarian essayist and farmer Gene Logsdon exemplified this view, and yet even he sensed the dangers inherent in upholding the inward-looking Amish as a model for others. "In this age," he asked us, "can we, dare we, upraise and canonize communities who cut themselves off that much from the rest of the population?" Others too have pointed out that relatively homogeneous, small-scale societies like the Amish often embody the positive virtues of community and localism at the very same time that they reflect a relative intolerance for diversity and inclusiveness.[47]

The misperceptions held by non-Amish can be difficult to dislodge, however, because they are psychologically and economically convenient.[48] For those who want to be environmentally ethical consumers, the shorthand of the ecological or the noble Amish is appealing precisely because in the wider society the conceptual category "sustainable natural environment" is not easily made legible, visible, or recognizable.[49] It takes a lot of time and effort to keep track of which fish are sustainably caught, which hardwoods are sustainably harvested, or which vegetables are sustainably grown. The visibility of the Amish seems to offer a way out of this conundrum, because a product associated with the image of a buggy or a hat or even the Amish name conjures up the assurance of sustainability. But stereotypes about the Cultural Other, even flattering ones, ultimately do a disservice and tell us more about the observer than about those to whom the labels are applied.

The ecological attitude that has been attributed to them by outsiders has been internalized by some Amish as a point of pride and a mechanism of articulating their "difference" from the larger society. Such images may provide not only an economic benefit but also a sense of empowerment. Like some indigenous peoples, Amish who come to see themselves as all natural "demonstrate to themselves and to the world that their traditions, far from being obsolete, and out of touch with modern reality, express a truth of urgent relevance for the future of the Earth."[50] The stereotypes can thus serve as a positive motivating force. "With the tourists and everybody watching, we want to be an example," said one New Order man. Adopting such images of themselves is risky, however, because when the Amish do embrace change, it can undermine their credibility and open them up to charges of hypocrisy.

In the end, however, we agree with those scholars who argue for a middle ground, humanizing the Cultural Other so that they are neither saints nor demons but people with virtues and faults like anyone else.[51] Achieving this middle ground and forging meaningful partnerships requires that each side meet the other halfway. It requires those who do not share the Amish theological and cultural orientation to the world to recognize their own blinders, prejudices, and predispositions. In some cases, outsiders simply express indifference to Amish opinions and lifestyle choices. In other cases, the absolutism and missionary zeal of some environmentalists and animal-rights advocates lead them to categorically dismiss Amish points of view. On the Amish side, the middle ground requires a willingness to adopt a more flexible perspective on their place in the world and to recognize the permeability of the boundaries between themselves and their non-Amish neighbors. The Amish have every right to live apart, but when they engage in businesses or activities that impact others and affect the public good, they must be willing to acknowledge the environmental responsibilities that the larger world demands. Many examples of cooperation with public officials, from immunization campaigns and road safety to health insurance and schooling, already exist and could serve as a model for such efforts.

In the final analysis, our case study of the Amish leads us back to a critique of our own society and its resource-degrading character. The fact that Amish lifestyles are increasingly incorporating elements of modernity is not a criticism of the Amish themselves so much as a reminder that we all contribute daily, in ways large and small, to the ecological degradation of the earth. The Amish may be parochial in their orientation to the world, but they *do* see themselves as stewards of the earth, whether or not they emphasize dominion. By and large they resist the mind-set of thoughtless exploitation and remain opposed to urban and suburban sprawl and to wholesale destruction of rural landscapes. In this sense, dispelling the myth that the Amish have a "natural propensity" for living sustainably should not be equated with a pessimistic environmental message. Quite the opposite. It means that non-Amish must try harder to understand Amish culture in its own context, to be aware of the consequences of interpreting Amish actions out of context, and to see Amish approaches to nature as a potential catalyst for self-reflection.

Methods

We used a variety of methods to gather information about Amish engagement with the natural world in order to explore attitudes and behaviors in different social contexts. Over the course of our seven years of fieldwork we spoke at length with approximately 150 Amish and 30 non-Amish individuals in numerous states. Since our home base, the College of Wooster, is on the northern edge of the Holmes County settlement in Ohio, we conducted approximately 65 percent of our interviews and observations in this area. Our proximity to the Holmes County settlement allowed us to get to know one Amish-populated area extremely well. The diversity of the Holmes County settlement presented a natural laboratory for observing similarities and differences between the ways groups across the Amish spectrum engaged with nature. At the same time, we realize that this settlement's overall level of affluence, as well as its size and diversity, are not completely representative of Amish life. Fully one-half of Amish settlements are made up of only one church district, and these are often located in rural areas far from major population centers. In addition, ecological landscapes, and the opportunities and constraints they impose, differ from region to region, as do the environmental regulations that have been enacted in different states and localities.

For these reasons, we made every effort to visit settlements large and small in other areas of Ohio and in other states, including outside the Midwest. We made four trips to Lancaster, Pennsylvania, and two trips to Elkhart-LaGrange, Indiana, each spanning several days. Our relative proximity to the Geauga settlement allowed us to occasionally attend events and to speak with Amish in the Middlefield, Ohio, area. In the summer of 2016 we took a three-week trip to Amish settlements out West, including Hulett, Wyoming, St. Ignatius and Rexford in Montana, Westcliffe and Monte Vista in Colorado, and settlements in Salmon, Idaho, and Arthur, Illinois. We visited the Big Valley area of Pennsylvania, the Pearisburg settlement in the Appalachian Mountains of Virginia, Crab Orchard, Kentucky, Hillsdale and Branch Counties in Michigan, and Chautauqua County, New York. We also made a concerted effort to visit some of the more isolated or conservative Amish areas in Ohio, including multiple visits to Swartzentruber settlements in Lodi and in Peebles; to Lakeville–Big Prairie, home of a conservative branch of Andy Weaver Amish;

and to settlements at West Union in Adams County and Glenmont-Brinkhaven in Knox County.

Interviews

Our formal interviews included an initial sample of approximately fifty individuals who had indicated on our survey that they were willing to meet with us, as well as a snowball sample of another hundred individuals obtained by asking each interviewee to recommend knowledgeable people on a given topic. Most of our interviews occurred in family homes, sitting around the kitchen table or in the living room, but sometimes we sat outside, in a shop or office, or in the barn. The conversations typically lasted one to two hours. We took detailed, handwritten notes of all conversations, and we digitally recorded approximately 60 percent of our interviews, including most of those with Old and New Order participants. We did not attempt to tape-record conversations with Amish from conservative affiliations, because we felt that it would make them less open to sharing their thoughts. A small group of dedicated student assistants transcribed our tape-recorded interviews.

In addition to speaking with Amish individuals, we interviewed about twenty former Amish in multiple states who had grown up in diverse affiliations ranging from Swartzentruber and Swiss Amish to Old and New Order. They had unique perspectives on our topic because they had lived an Amish lifestyle and yet had achieved some critical distance from it. To obtain yet another outsider perspective on Amish engagements with nature, we interviewed approximately thirty non-Amish individuals who work closely with the Amish in various capacities. Agricultural-extension agents and soil and water conservation district (SWCD) officials contributed their perspectives on Amish farming practices and runoff. Department of Natural Resources wildlife officials spoke with us about Amish hunting, fishing, and trapping. Forestry professionals shared their insights into Amish timber cutting. And the list goes on: veterinarians, milk inspectors, tour-bus drivers, health commissioners, local newspaper reporters, medical professionals, and realtors. By triangulating the varied perspectives of Amish, former Amish, and non-Amish, we hoped to get a clearer, more nuanced understanding of our topic.

Observation and Participant Observation

While meeting with Amish individuals in their homes and shops, we had many opportunities to chat informally and to observe spontaneous interactions with animals and with the landscape. We arrived at a Swartzentruber home one morning as the husband and wife were butchering dozens of chickens and watched as they dipped the birds in boiling water and then put them in a gasoline-powered "plucker" for defeathering. On a visit to an Old Order family we were shown several antler mounts, as well as photos of the children and the trout they had caught on a camping trip to a wilderness area. We arrived at another Swartzentruber home just after the father had dispatched a beef cow with a .22 magnum and hung it from the barn door. "Do you notice anything?" he asked. "We forgot to take one of the hooves off! Oh well, we'll just mix it in with the beef," he laughed.

On many occasions, our Amish hosts gave us guided tours of their businesses, allowing us to ask questions while observing, for instance, a sawmill in full operation or a row of stalls for stallions at a breeding barn. Where possible, we attended events that were planned and attended primarily by the Amish community but also open to the general public, such as Horse Progress Days and Family Farm Field Days. We also frequented events that attracted many Amish vendors and attendees, such as home and garden shows, health fairs, exotic-animal auctions, and sportsman shows. In a few cases, we were invited to special Amish-organized events that had particular relevance for our topic. These included benefit trail rides, parochial-school classes, a family outreach seminar for parents whose children suffered from propionic acidemia, bird and butterfly counts, a meeting for organic-produce growers, and a trapper's licensing workshop for young Amish boys. Our trip with an Amish family to Texas and Arizona provided a rare opportunity to spend a sustained period of time traveling and exploring the outdoors together. On our trip to Montana we were fortunate to be hosted by an Amish family and attend Sunday school and a singing in their church district.

Documents

In addition to drawing on secondary scholarly sources, we filled an entire bookshelf with primary documents. The growth in recent years of Amish-authored publications has provided a wealth of printed material that is accessible to outsiders. We used these as a window into Amish views on nature in several ways. We subscribed to or collected back issues of some of the more general correspondence newspapers, regional newsletters, and religious periodicals—*Plain Interest, Family Life, Blackboard Bulletin, Young Companion, Gemeinde Register, Keepers at Home, Connections*—and perused them for nature-themed material. We also subscribed to or collected back issues of nature-themed topical magazines, such as *Farming, Nature Trails, Bobolink, Truck Patch News, Feathers and Friends*, and *Hometown Outdoors*. We bought and read numerous Amish-authored books on topics such as big-game hunting, birding, cooking, and herbal remedies. Whenever we visited a business or an event, we always picked up brochures, fliers, calling cards, and other printed material. Collectively, these provided a wealth of visual images and vernacular accounts of Amish constructions of nature.

Questionnaire Survey and Carbon-Footprint Calculations

In order to compare Amish and English ecological footprints, we designed and sent out a household-resource-use survey. Our survey document was three pages long, double-sided, and focused on household expenditures, diet, and resource use. We asked respondents who also operated a business out of the home to report only expenses related to their home and household. While resource use in Amish businesses deserves exploration, we limited our survey to households in order to keep our data set manageable.

In an effort to encourage our survey recipients to respond, we made the survey as welcoming as possible. We explained our goals, illustrated the pages with line drawings, hand-addressed the envelopes, and wrote a message by hand on each survey we sent out. A week before the survey was mailed, the *Budget*, a Sugarcreek, Ohio–based newspaper widely read by the Amish, ran a front-page article announcing the survey and explaining our goals. We included a self-addressed, stamped return envelope with each survey, as well as a photocopy of the *Budget* article. Surveys were anonymous, but we asked for contact information from respondents who wanted to receive a summary of the results or were willing to participate in a follow-up interview. However, we did not use the identification data in any way during data analysis. We presented some survey questions in a different order for each group so that we could identify the affiliation of each respondent without other identification. We tested the survey on English and Amish friends during development and tried to refine the questions to make them easy to understand and answer.

We sent out 1,500 surveys, 800 to Amish and 700 to English recipients. The surveys were mailed in August 2015 and requested information for the year preceding the mailing date. The number of Amish recipients we contacted from the various affiliations was based on their proportion in the Holmes County settlement. Using the Amish directory for the Holmes County settlement, we sent the survey to every tenth or thirteenth family, for example, depending on the number of households and the number of surveys allotted for that affiliation. We sent 400 surveys to Old Order, 250 to Andy Weaver, and 150 to New Order families. We identified English recipients using the local telephone directory. When we came across typical Amish surnames, we checked the directory to be sure that the family was not Amish. We also chose addresses that were in the southern portion of the area covered by the local telephone directory, which overlapped geographically with the Amish settlement.

We received 41 returned surveys from Andy Weaver families (16.4 percent response rate), 76 from Old Order families (19 percent), and 37 from New Order families (24.7 percent). The return rate from English families was 30.6 percent (214 surveys). Responses were entered into an Excel file and analyzed using Excel (for means, SE, and counts) and SPSS (for statistical analyses).[1] Not all respondents answered every question. In cases where surveys lacked energy-expenditure data and these data could not be estimated, we eliminated that survey from the carbon-footprint analysis but used it for other analyses. Of our returned surveys, a total of 155 Amish (94.5 percent) and 139 non-Amish (65.2 percent) surveys were entered into a footprint calculator.

Where a household indicated that it used a particular fuel source but did not specify either the volume or the cost of that fuel, we substituted the average value for other surveys in that affiliation. We made similar substitutions of group averages for other data on a case-by-case basis and only where a very small fraction of the surveys lacked that information. Where respondents gave us the costs (in dollars) of their annual energy bills, we converted those costs to volumes (such as kilowatt-hours of electricity or cubic feet of natural gas) using the average cost per volume of that fuel in Ohio in 2014 (the winter of the year prior to the survey) taken from internet sources.[2]

The online carbon calculator expressed carbon footprints in terms of metric tons of carbon dioxide equivalents, including actual carbon dioxide emissions as well as the warming potential generated by other greenhouse gases, principally methane and nitrous oxide. Methane's global-warming potential is 25 times that of carbon dioxide, and nitrous oxide's is 298 times that of carbon dioxide. Emissions of these other greenhouse gases were multiplied by their warming potentials to calculate their carbon dioxide equivalents (CO_2e).

The calculator did not include horse-based transportation. We estimated the carbon cost of a buggy horse by calculating its carbon emissions from (1) foodstocks to feed a horse for one year, (2) enteric emissions from food consumption per year, and (3) greenhouse-gas production from manure management. We assumed that an average horse consumes 12.5 pounds of hay or forage a day, or 4,380 pounds (1,987 kg) per year. We assumed that in Ohio five acres, or 2.023 hectares, of pasture were required per horse.[3] We assumed that Amish would be using manure as fertilizer for that pasture, with occasional inputs of lime. We used an online calculator, the Farm Energy Analysis Tool (FEAT) from Pennsylvania State University, to estimate carbon emissions for three different pasture vegetations (clover, alfalfa, and switchgrass) and averaged the three values, since a pasture is likely to be a mix of grass and legumes.[4] This calculator allowed us to specify that the pasture did not receive fossil-fuel inputs. We confirmed the reasonableness of our value by comparing it with values in other sources that gave data for carbon from hay or grazing land.[5] Emissions from digestive processes and from manure management for horses were estimated using published tables for agricultural greenhouse-gas production.[6] We determined that the average carbon output per horse was 1,458.2 kilograms carbon dioxide equivalents per year, or 1.45 metric tons of carbon dioxide equivalents. We determined that driving 2,490 automobile miles (at 22 miles per gallon) would generate 1.45 metric tons of carbon dioxide equivalents in the calculator, and we used this figure to approximate the carbon footprint of each horse.

In the data tables, we averaged transit miles traveled by different vehicles using the total miles reported by each household. English miles were assigned directly to that household since they were driven in a private vehicle. Some Amish van travel, however, is shared with other households. To account for this ride-sharing, we discounted each Amish family's van miles by one-third in calculating carbon emissions. We estimated that about half of both short- and long-distance van trips were made by family members alone (for errands, medical appointments, to visit family, for leisure travel) and that the remainder were shared with people from more than one family. So the group trips (half the miles) were divided by three, assuming that they were shared with two additional households. The actual values will differ from family to family.

The online calculator accepted electric, natural gas, and liquid-petroleum fuels as inputs but did not include the carbon output from burning wood. We calculated the average greenhouse-gas emissions of burning wood using data for one cord of oak, estimated to weigh 4,200 pounds. Carbon emissions for wood were taken from the Environmental Protection Agency emissions hub.[7] We did not include carbon contributions from cutting, sawing, or transporting the wood. Based on our calculations, one cord of wood produces 3.5 metric tons of CO_2e, which, when entered into the calculator as 235 gallons of heating oil, gave the correct carbon output. Thus, we substituted heating oil for each cord of wood used by respondents. Many respondents told us how much they paid for wood but not the volume they purchased. We used different conversions for Amish and English wood costs, since those surveys that did report both volume and cost gave different average values. For Amish, we estimate fifty-four dollars per cord; for English, eighty dollars per cord.

TABLE A.1.
Diet estimates from the Cool Climate Calculator

Food category	Estimated daily caloric intake for Ohio[a]	Average servings per day
Meat, poultry, fish	543	2
Dairy	286	2
Fruits and vegetables	271	5
Grains and baked goods	669	4
Other (snacks, chips, candy, soft drinks, alcohol)	736	5

[a]Values are for each individual in a family of average size and income in Ohio.

The online calculator asked respondents to estimate their caloric intake against the average daily caloric intake for various kinds of foods (meat, dairy, fruits and vegetables, grains and breads, and snacks and beverages). We translated calories into an estimate of "servings per day."[8] We gave respondents a scale ranging from "none" through "average" and up to "three times average" and listed the number of servings of each food type that would be "average" based on calorie values. We could then enter each respondent's relative score for each food category into the online footprint website. The food categories are shown in table A.1.

A similar method was used to assess consumption (purchases and spending) by households. The online calculator identified fifteen different categories of spending and provided the average amount spent by a "typical" household on each category per month. To make the survey less onerous, we collapsed these fifteen categories into seven categories of goods and services. We then asked respondents to rate their family spending relative to the average. The consumer categories we used are shown in table A.2. We calculated two overall consumer indexes, one for goods and one for services, families' proportional spending for each category weighted by the monetary value of that category. Virtually every respondent provided us with information on spending habits. However, the broad categories we used to make the survey user-friendly also limited our ability to make more detailed comparisons of respondents' consumption.

We entered each complete survey file into the online carbon calculator. Because our survey was anonymous, we used Ohio data as the baseline and used "average" as the income choice for every survey. The location and income of the user is not part of carbon-footprint calculations but generates a baseline against which each

TABLE A.2.
Categories of goods and services used to assess consumer carbon footprint

Consumer Item	Average spending per month ($)
Goods	
Clothing and shoes	282
Furniture and appliances	297
Other (books, personal-care products, cleaning, auto parts, entertainment)	485
Services	
Medical appointments, health care	781
Vehicle and household maintenance	172
Cable, internet, cell phone, finance, education	619
Organizations, charity, miscellaneous	222

Note: The average monthly spending values are defaults calculated by the Cool Climate Calculator based on an average family size and average household income in Wooster, Ohio. Respondents indicated how their expenditures compared with this average value.

user's efficiency can be measured. We do not report efficiencies; we simply report the carbon emissions generated by each household.

We used SPSS statistical software to test hypotheses about the data set. We tested data for each affiliation and each variable for normality. Where sample sizes were relatively large and data were mostly normal (a minority of cases) we used an analysis of variance, followed by Tukey's post-hoc test, to assign significant differences. For non-normal data, we analyzed and compared survey data and total carbon data using a Kruskal-Wallis test, followed by individual Mann-Whitney U tests to identify homogeneous groups.

There were statistical challenges in our data set that sometimes made it difficult to identify differences between our study populations. Demographic characteristics of Amish and English communities make it inherently difficult to collect samples that are balanced for household size. A random sample of English households will not yield enough homes with eight or ten children to balance the sample size of that group within the Amish community. Conversely, two-person Amish families (grandparents) seldom live independently, but rather in a Dawdi Haus arrangement,[9] sharing much of their home economy with a married child and grandchildren. With very unequal sample sizes, such as those for our carbon-footprint analysis, the power of any statistical test is diminished and the probability of a type II error is increased. This means that it is harder to detect differences between groups. Small sample sizes also have higher variances and are less likely to accurately portray the groups of which they are a part. This can also obscure differences we might hypothesize to be present in the data. We can see trends, but we lack the statistical power to actually conclude differences among categories.

Our study also included the New Ecological Paradigm (NEP) survey to assess environmental attitudes among our respondents. This fifteen-question survey has been used among different ethnic groups and in different nations to measure beliefs and attitudes that might inform environmental concern.[10] Respondents are asked to score each question on a five-point Likert scale, ranging from "strongly agree" to "strongly disagree." Where appropriate, we reverse-scored questions so that the magnitude of a score was always a reflection of a similar attitudinal tendency.

The wording of survey questions is important, since phrases might be understood differently by people with different vocabularies or points of reference. This could complicate expected patterns of intercorrelation among the questions on the survey or among the three questions making up each concept group. We evaluated scale reliability of the NEP scores using Cronbach's alpha on the fifteen survey questions. English surveys (Cronbach's alpha = 0.864) showed strong scale reliability; Amish surveys (alpha = 0.674) were lower but not problematic. We also tested each of the five concept groups for reliability among the three questions in each group. Here, four of the five English Cronbach's alpha values were greater than 0.500 (range 0.464 to 0.795). However, four of five Amish values were under 0.500 (range 0.136 to 0.625). This suggests that our Amish survey respondents interpreted some of these questions differently than their English neighbors did. The higher Cronbach's alpha values suggest that our English respondents gave similar answers to all three questions in each concept group, implying that the three clustered questions all asked about the same ideas. For the Amish, however, in several cases low Cronbach's alpha values indicated that questions assumed to address similar concepts were being interpreted as having somewhat contradictory meanings. We tried to account for that in discussing our results.

We analyzed the actual NEP scores for individual questions and groups of questions using an analysis of variance. Where the analysis of variance was significant, we used Tukey's post-hoc test to determine which variables differed from one another. We explored our data using a second-order factor structure to examine particular conceptual elements of environmental attitudes that might differentiate our study populations.[11]

Preface

1. Louv, *Last Child in the Woods.*

2. Taborri I. Bruhl, "The Amish Question," *Sustainability Us* (blog), 29 May 2013, http://sustainableus.org/2013/05/29/the-amish-question/.

3. Olshan, "What Good Are the Amish?," 242.

4. See Donald Kraybill, "Fake Amish and the Real Ones," *Huffington Post*, 18 July 2013, for a debunking of these and other myths.

5. The World Commission on Environment and Development of the United Nations, in the 1987 Brundtland Report, *Our Common Future*, defines sustainable development as "development that meets the needs of the present without compromising the ability of future generations to meet their own needs." See http://www.un-documents.net/our-common-future.pdf.

6. Netting, *Smallholders, Householders*, 21.

7. An entire interdisciplinary field of study has emerged in recent years to engage in "critical inquiry into the relationships among human beings and their diverse cultures, environments, religious beliefs and practices." See International Society for the Study of Religion, Nature and Culture, https://www.issrnc.org/.

8. Sponsel, *Spiritual Ecology*, 47.

9. Sponsel, *Spiritual Ecology*, 47.

10. See Bourne, *End of Plenty*, for an account of the relationship between the global food supply and population growth; and Kolbert, *Sixth Extinction*, for an analysis of how the current era of species extinctions compares with five previous epochs.

11. Waters et al., "Anthropocene," argues that over the past century the appearance of manufactured materials in sediments, the modification of carbon, nitrogen, and phosphorus cycles, the rates of sea-level rise, and the extent of human perturbation of the climate system, as well as species invasions worldwide and accelerating rates of extinction, "render the Anthropocene stratigraphically distinct from the Holocene and earlier epochs" (aad2622).

12. *Gemeinde Register*, 7 Aug. 2013, 20.

13. Kraybill, Johnson-Weiner, and Nolt, *Amish*, 15.

14. Kraybill, Johnson-Weiner, and Nolt, *Amish*, 5. Settlements may include church districts from multiple affiliations. But approximately half of all Amish settlements comprise only one or two church districts and represent just one affiliation.

15. Olshan, "What Good Are the Amish?," 236, makes a similar point, arguing that the assumption that the Amish live in some pristine, isolated "Amishland" is an inadequate basis for admiration.

CHAPTER 1: Deciphering the Amish Relationship with Nature

1. Sleeth, *Almost Amish*, xvii.

2. Sleeth, *Almost Amish*, 71.

3. Reiheld, "Donald G. Beam," 242.

4. Reiheld, "Donald G. Beam," 246.

5. Interview by David McConnell, Wooster, OH, 2 Aug. 2012.

6. Weaver-Zercher, *Amish in the American Imagination*, 185. See also Olshan, "What Good Are the Amish?," 231.

7. We use the phrase "ecological Amish" to capture the popular image of the Amish as ecologically sensitive and in tune with nature, in the same way that Shepherd Krech III uses the book title *The Ecological Indian* to invoke the relation between indigeneity and ecology in the American public imagination.

8. Reschly, "Midwestern Amish since 1945," 291.
9. Krech, *Ecological Indian*, 21. Of course, many differences exist between the Amish and indigenous Americans, not the least of which is the racial and ethnic context of persecution.
10. Berry, *Bringing It to the Table*, 112.
11. Lopez, *Rediscovery of North America*, 51.
12. Kline, *Great Possessions*, xix. Kingsolver devotes part of chapter 11 in her *Animal, Vegetable, Miracle* to reflections on her visit with the Kline family.
13. Hostetler and Huntington, *Amish Children*, 114.
14. Hostetler and Huntington, *Amish Children*, 7.
15. Redekop, *Creation and the Environment*, argues that a strong concern for environmental issues is an integral part of Anabaptist tradition.
16. Sindya N. Bhanoo, "Amish Farming Draws Rare Government Scrutiny," *New York Times*, 8 June 2010.
17. Jimmy Doherty, "Amish Farmers Embrace GM Crops," *BBC News*, 24 Nov. 2008, accessed 15 Nov. 2017, http://news.bbc.co.uk/2/hi/science/nature/7742471.stm.
18. Blake et al., "Modern Amish Farming as Ecological Agriculture."
19. Vonk, *Sustainability and Quality of Life*, 94–95.
20. Loewen, "Quiet on the Land," 153. Loewen refers to this tendency as "an Anabaptist exceptionalism."
21. Wexler, *When God Isn't Green*, 11–39.
22. White, "Historical Roots of Our Ecologic Crisis," 1205.
23. White, "Historical Roots of Our Ecologic Crisis," 1205.
24. Nash, *Wilderness and the American Mind*, 9.
25. Taylor, Van Wieren, and Zaleha, "Lynn White Jr.," 1000. See also Taylor, "Wilderness, Spirituality, and Biodiversity."
26. Kraybill, Johnson-Weiner, and Nolt, *Amish*, 26.
27. Beachy, *Unser Leit*, 1:375–76.
28. According to an Amish man in Big Valley, Pennsylvania, the early settlers chose the Blue Mountains because the area reminded them of home.
29. Nolt, *History of the Amish*, 82.
30. "Amish Population Profile, 2017," Amish Studies, http://groups.etown.edu/amishstudies/statistics/amish-population-profile-2017/.
31. Donnermeyer, Anderson, and Cooksey, "Amish Population," 74.
32. Kraybill, Johnson-Weiner, and Nolt, *Amish*, 282, provides data from settlement directories from 2006 to 2010 that show the percentage of Amish household heads whose primary occupation was farming ranging from 36 percent in Lancaster and 17 percent in Holmes County to 14 percent in Elkhart-LaGrange and 7 percent in Geauga. But their figures include animal husbandry and greenhouse management, occupations very different from typical farming.
33. For an in-depth discussion of the reasons behind the successful transition from farming to market-oriented businesses, see Kraybill and Nolt, *Amish Enterprise*; and Wesner, *Success Made Simple*.
34. See Moledina et al., "Amish Economic Transformations," for a detailed analysis of social-class differentiation in one Ohio community.
35. Kraybill, Johnson-Weiner, and Nolt, *Amish*, 60.
36. Genesis 9:12–13.
37. Hebrews 11:13.
38. Kraybill, Nolt, and Weaver-Zercher, *Amish Way*, 139.
39. Kraybill, Nolt, and Weaver-Zercher, *Amish Way*, 139.

40. Taylor and Buttel, "How Do We Know We Have Global Environmental Problems?," 407.

41. Kraybill, Johnson-Weiner, and Nolt, *Amish*, 73.

42. Hurst and McConnell, *Amish Paradox*, 34.

43. Kraybill, Johnson-Weiner, and Nolt, *Amish*, 312.

44. Kraybill, Johnson-Weiner, and Nolt, *Amish*, 24.

45. Robbins, *Political Ecology*, 11–24. Biersack, "Reimagining Political Ecology," 3, notes that the neo-Marxist scholar Eric Wolf was the first to coin the term "political ecology," referring to the study of "how power relations mediate human-environment relations."

46. Biersack, "Reimagining Political Ecology," 5, describes the terrain of political ecology as comprising culture, power, history, and nature, a framework we find useful for our inquiry.

47. Conklin, "Study of Shifting Cultivation," 60. Virginia Nazarea's *Ethnoecology* is a contemporary application of Conklin's approach that focuses on situated knowledge.

48. The literature on gender roles among the Amish is extensive. For overviews, see Olshan and Schmidt, "Amish Women and the Feminist Conundrum"; and Kraybill, Johnson-Weiner, and Nolt, *Amish*, 192–211.

49. Nolt, "Who Are the Real Amish?," 385.

50. Billig and Zook, "Functionalist Problem," for example, argues that most scholarly work on the Amish falls into a functionalist framework that overstates the harmonious interplay between values, rituals, and social structures and by implication, de-emphasizes conflict, power, inequality, and dysfunction.

51. Ching and Creed, "Recognizing Rusticity," 5.

52. Kraybill, Nolt, and Weaver-Zercher, *Amish Way*, 138.

53. Ching and Creed, "Recognizing Rusticity," 4, argues that place is a central axis of identity and a basic category of human experience, much like age, gender, race or ethnicity, social class, and sexual orientation.

54. Crumley, "Historical Ecology," 7, argues that an ecological analysis of landscapes involves "knowledge systems, productive practices, and religious rites that local peoples have developed over the course of centuries as a means of interacting with and gaining sustenance from their biophysical environments."

55. Lefebvre, quoted in Sheridan, *Landscapes of Fraud*, 6.

56. See Kraybill, Nolt, and Weaver-Zercher, *Amish Way*; and Kraybill, Johnson-Weiner, and Nolt, *Amish*, 97–114.

57. Bailey, Jensen, and Ransom, "Rural America in a Globalizing World," xiv.

58. Pinchot, *Fight for Conservation*, 40.

59. Muir, *Yosemite*, 262.

60. See Berry, *Devoted to Nature*, 79–81, for a discussion of the scholarly debate over Muir's theological views. We are grateful to Steve Nolt for pointing out the parallels between Amish views of nature and early-twentieth-century conservationist approaches.

61. See Kraybill, Johnson-Weiner, and Nolt, *Amish*, 145–53, for an extended discussion of the causes of church schism, which include migration history, theological disagreements over how to define separation from the world, and the personalities and styles of church leaders.

62. Kraybill, Johnson-Weiner, and Nolt, *Amish*, 140.

63. See Kraybill, Johnson-Weiner, and Nolt, *Amish*, 139–42, for a list of the forty Amish affiliations in North America and for a discussion of the origins of the names for select affiliations.

64. Beachy, *Unser Leit*, 2:471.

65. Kraybill, Johnson-Weiner, and Nolt, *Amish*, 138.

66. Hurst and McConnell, *Amish Paradox*, 35.

67. We recognize that this shorthand creates a somewhat oversimplified portrait and will point out exceptions and caveats when they are relevant.

68. The sociologist Ulrich Beck, for instance, writes, "The only question is: how do we handle nature *after* it ends?" Beck, "Ecological Questions," 270.

69. See Crutzen, "Geology of Mankind," for a concise synopsis of the term "Anthropocene."

70. For scientific analyses of human impacts on the global ecosystem, see Sanderson et al., "Human Footprint"; and Vitousek et al., "Human Domination of Earth's Ecosystems."

71. Biersack, "Reimagining Political Ecology," 4.

72. Cronon, "Trouble With Wilderness," 19. Dove and Carpenter, "Introduction," also questions this dichotomy.

73. We concur with Little, "Environments and Environmentalisms," 257, which argues that "biological and geological processes cannot be subsumed under discourse theory, just as political and cultural change cannot be subsumed under the concept of natural selection."

74. Milton, *Environmentalism and Cultural Theory*, 215.

75. Rappaport, *Ecology, Meaning, and Religion*, 97.

76. Parajuli, "How Can Four Trees Make a Jungle?," 17.

77. Milton, "Cultural Theory and Environmentalism," 251.

CHAPTER 2: Raising Children at Nature's Doorstep

1. See, e.g., Alexandra Jardine, "Nature Valley Guilt-Trips Parents over Their Screen-Addicted Kids," 23 July 2015, http://creativity-online.com/work/nature-valley-rediscover-nature/42824.

2. Rhonda Clements surveyed more than eight hundred mothers nationwide and found that children today play outdoors less frequently than the previous generation (70 percent of mothers reported playing outdoors every day, compared with only 31 percent of their children). In addition, increased use of television and computers in the home (up to four hours a day) was found to be a major obstacle to outdoor play. Clements, "Investigation of the Status of Outdoor Play," 72–74. See also Kimball, Schuhmann, and Brown, "More Kids in the Woods," 373.

3. Berry, *Dream of the Earth*, argues that our fascination with gadgets is a result of our society being caught in a "closed cycle of production and consumption" that, in turn, rests on an "industrial-technological fundamentalism" (57).

4. David Kline, "Letter from Larksong," *Farming Magazine*, Spring 2007, 7.

5. Kraybill, Nolt, and Weaver-Zercher, *Amish Way*, 99.

6. Orr, "Political Economy," 291. Orr also sees the Amish as exceptions to the move toward isolated individualism, increased violence, a faster pace of life, and immersion in virtual reality.

7. In a few of the larger or older settlements, public schools are an option chosen by a small minority of parents. See McConnell and Hurst, "No 'Rip Van Winkles' Here," for a discussion of public-school attendance in the Holmes County Settlement in Ohio.

8. Taylor and Buttel, "How Do We Know We Have Global Environmental Problems?," 407.

9. See Johnson-Weiner, *Train Up a Child*; and DeWalt, *Amish Education in the United States and Canada*.

10. Ediger, "Teaching Science in the Old Order Amish School."

11. Anderson, "Amish Education," 6.

12. Milton, *Loving Nature*, 13, argues that science involves "the systematic search for knowledge, characterized by induction (verification through observation) and reduction (explanation of phenomena in terms of their progressively smaller components)." It is also open-ended, generates new knowledge, employs a rigorous methodology, and "has to obey rules which do not constrain common sense."

13. Institute in Basic Life Principles, *Character Sketches*, 5.

14. Johnson-Weiner, *Train Up a Child*, 58 and 70.

15. Johnson-Weiner, *Train Up a Child*, 79.

16. A Beka, a Florida-based home-schooling company whose geography and history textbooks are widely used in the northern Indian schools, also publishes a series of elementary-school science texts.

17. Some Amish schools in northern Indiana and Iowa do participate in standardized testing of mathematics and English, and Amish students generally achieve at levels comparable to those of their non-Amish counterparts in the public schools. Yet Amish schools that participate are on the progressive end of the spectrum, so extrapolating these results to the broader Amish population is dangerous. See Johnson-Weiner, "Old Order Amish Education," 34.

18. In 2015, American fifteen-year-olds taking the cross-national test known as the Program for International Student Assessment ranked nineteenth out of the thirty-five countries that are members of the Organization for Cooperation and Economic Development. See Drew Desilver, "U.S. Students' Academic Achievement Still Lags That of Their Peers in Many Other Countries," Pew Research Center, 15 Feb. 2017, http://www.pewresearch.org/fact-tank/2017/02/15/u-s-students-internationally-math-science/.

CHAPTER 3: The Amish Ecological Footprint

1. See MacLeish, "If Life Means Going Without," for an interpretation of the Amish as a historical anachronism. Olshan, "Modernity, the Folk Society, and the Old Order Amish" offers one of the earliest critiques of the notion of the Amish as frozen in time.

2. Bill McKibben, "A Special Moment in History," *Atlantic*, May 1998.

3. Life cycle analysis (LCA) is an attempt to account for all the environmental consequences of a product, from its fabrication to its ultimate disposal as waste. See Ayres, "Life Cycle Analysis."

4. Ryan and Durning, *Stuff*.

5. Wackernagel and Rees, *Our Ecological Footprint*.

6. Vonk, *Sustainability and Quality of Life*, 81–82; Bender, *Plain and Simple*.

7. For more details about how we conducted the survey, see the appendix.

8. In this chapter, we use the adjective "significant" only when the differences are *statistically* significant, at a significance level (alpha) of 0.05.

9. Sample sizes among groups were not equal, but data were relatively normally distributed. Variances were not similar (Levene's test, $p < 0.001$), so we used the Welch's robust analysis of variance. Tukey's post-hoc test was used to detect differences among groups.

10. English respondents owned from one to four automobiles; the median number was two. For gas mileage, $n = 213$, mean = 23.47 mpg, SE = 0.3180.

11. Fossil-fuel neutral means that the mode of transportation does not use geological fossil deposits (oil, natural gas); rather, horses metabolize the organic carbon-based molecules synthesized by pasture plants during photosynthesis. But cultivating these plants and their metabolism by the horse generate wastes, including carbon dioxide and other molecules, which enter the atmosphere and have an environmental effect over the short term.

12. Similar outputs come from pets, but we did not assess the environmental costs of pet ownership, because the situations surrounding their care varied. Nor did we assess the carbon output of draft horses or other farm animals that were part of household businesses.

13. Carbon dioxide is the most abundant greenhouse gas, but other atmospheric gases, especially methane and nitrous oxide, also trap heat and cause atmospheric warming.

14. It is not unusual for Amish families to travel to Mexico for some health procedures.

15. Some New Order churches permit grid electric power, although this is rare in the Holmes County settlement. Other Amish occasionally have grid power for special reasons, particularly if needed to power medical equipment.

16. Andy Weaver and Swartzentruber Amish are generally prevented from using solar power by their Ordnung.

17. Coal, natural gas, and petroleum are the organic (carbon-containing) remains of ancient plants that have been buried and transformed by fossilization.

18. Partnership for Policy Integrity, Carbon Emissions from Burning Biomass for Energy, http://www.pfpi.net/carbon-emissions.

19. Hurst and McConnell, *Amish Paradox*, 107.

20. Only 5 out of 213 English respondents indicated that they did not eat meat or fish.

21. Some Old Order and New Order Amish are allowed to join with neighbors to build a freezer barn on a small plot of land and share the cost of electric power from the public utility grid to run their individual freezers.

22. Completely local food production would, however, reduce the greenhouse-gas-emission costs of household foods by only about 4 percent. While the concept of food miles is easily grasped, most of the greenhouse-gas emissions associated with food come from its production, not from its transportation. See Weber and Matthews, "Food-Miles and the Relative Climate Impacts," 3511–12.

23. The Cool Climate Network, University of California, Berkeley, http://coolclimate.berkeley.edu/index. Categories for the consumption measures are given in table A.2.

24. English households had higher expenditures than Amish households in the health-care, vehicle and household maintenance, and connectivity categories.

25. The levels of consumption by English households and Andy Weaver, Old Order, and New Order Amish households of clothing, household appliances and furniture, miscellaneous goods, and organizations and charities were not statistically different.

26. The Cool Climate Calculator is available at http://coolclimate.berkeley.edu/calculator.

27. Methane molecules have five times the warming potential of carbon dioxide, and nitrous oxide has 298 times the warming potential of carbon dioxide. To calculate their carbon equivalents, the volumes of methane and nitrous oxide must be multiplied by their global-warming-potential coefficients (5 and 298, respectively). When added together, these values give the overall warming effect of different greenhouse gases standardized to carbon dioxide measures.

28. For a detailed discussion of how we converted survey data into carbon calculations, see the appendix.

29. This is the baseline value predicted by the Cool Climate Calculator.

CHAPTER 4: The Transformation of Amish Agriculture

1. Weaver-Zercher, *Thrill of the Chaste*, 145.

2. Weaver-Zercher, *Thrill of the Chaste*, 148. See Marx, *Machine in the Garden*, for an analysis of sentimental pastoralism.

3. Pritchard, "Landscape Transformed," 12.

4. Reschly, *Amish on the Iowa Prairie*.

5. Kraybill, Johnson-Weiner, and Nolt, *Amish*, 281.

6. Kraybill and Nolt, *Amish Enterprise*, 35.

7. Stinner, Paoletti, and Stinner, "In Search of Traditional Farm Wisdom," 85.

8. Quoted from Monsanto's website, accessed 21 Feb. 2017, http://www.monsanto.com/whoweare/pages/default.aspx.

9. Quoted in Imhoff and Baumgartner, introduction, v.

10. Berry, *Bringing It to the Table*, 116, argues that Amish farming works precisely because it maintains a local economy that "steadfastly subordinates economic value to the values of religion and community."

11. Yoder, "Amish Agriculture in Iowa," 98–99.

12. Craumer, "Farm Productivity and Energy Efficiency," found central Pennsylvania Amish dairy farms to be 30–40 percent more energy efficient than non-Amish farms.

13. Jackson, "Amish Agriculture and No-Till," compared an Amish farm and a non-Amish farm in Holmes County, Ohio, and found that the Amish farm had soils with higher enzyme activity, more organic matter, and higher capacity to absorb water, characteristics usually associated with soil productivity.

14. David Kline, "Across the Editor's Desk," *Green Field Farms Newsletter* 9, no. 2 (June 2013): 1.

15. See Roc Morin, "The Amish Farmers Reinventing Organic Agriculture," *Atlantic*, 6 Oct. 2014.

16. Colony collapse disorder, first identified in 2007, occurs when workers in a hive suddenly disappear, leaving behind the queen and brood, without any clear provocation. Biologists suggest that colony collapse might result from several different causes, including parasites, pesticides, and poor nutrition. See https://www.epa.gov/pollinator-protection/colony-collapse-disorder.

17. When a hive gets too big, a group of worker bees, along with the old queen, leave in a "swarm" to find a new nest site.

18. Stated in the mission statement of the promotional flyer *Family Days on the Farm 2016*.

19. See Kraybill, "Plotting Social Change across Four Affiliations," 67.

20. Can milk does not have the same requirements for cooling as Grade A milk does and is usually reserved for the cheese market.

21. Johnson-Weiner, "Technological Diversity and Cultural Change," 7.

22. Kraybill, Johnson-Weiner, and Nolt, *Amish*, 283. Many Amish dairy farmers also rely on veterinarians to administer antibiotics to their animals, though according to Schewe and Brock, "Stewarding Dairy Herd Health," Amish dairy farmers used antibiotics less frequently than Mennonite and non-Anabaptist farmers and relied more on natural remedies for their animals.

23. See Ian T. Shearn, "Whose Side is the American Farm Bureau On?," *Nation*, 16 July 2012.

24. Steve Curwood, "Amish," *Living on Earth*, National Public Radio, 2 Dec. 1994, http://www.loe.org/shows/segments.html?programID=94-P13-00048&segmentID=1.

25. See Rick Marin, "From Gravy to Jus, Now 'Amish' is Trendy," *New York Times*, 17 Mar. 1999, http://www.nytimes.com/1999/03/17/dining/from-gravy-to-jus-now-amish-is-trendy.html.

26. Michael Pollan, "Playing God in the Garden," *New York Times Magazine*, 25 Oct. 1998.

27. Vonk, *Sustainability and Quality of Life*, 79, 81.

28. Berry, *Bringing It to the Table*, 106.

29. There are scattered instances of Amish farmers producing and selling organic products before the early 1990s.

30. "Poisoning the Earth," pt. 2, *Family Life*, May 1990, 29. The three-part series ran in *Family Life* from April to June 1990.

31. Under USDA rules, the land of farmers seeking the certified organic label must have been free of synthetic fertilizers, herbicides, and pesticide applications for three years. The lost income during this period is a major concern for conventional farmers considering a switch to organic farming.

32. The Amish case affirms Liz Carlisle's conclusion in "Audits and Agrarianism" that successful alternative food networks are governed primarily by moral economies and support networks and only secondarily by certification schemes.

33. See Peterson et al., "Motivation for Organic Grain Farming."

34. USDA, National Agricultural Statistics Service, "Organic Farming."

35. In Ohio, the USDA in 2014 listed 541 farms as certified organic, with $88.9 million in total sales, still just a fraction of the state's 75,000 farms and $10 billion in sales. See J. D. Malone, "More Ohio Farmers Go Organic," *Columbus Dispatch*, 15 Oct. 2015. But one Old Order organic farmer estimated that nearly half of those 541 farms were Amish.

36. Kraybill, Johnson-Weiner, and Nolt, *Amish*, 285.

37. Mariola and McConnell, "Shifting Landscape of Amish Agriculture." This section on Green Field Farms is a condensed and updated version of the earlier published article.

38. Green Field Farms website, http://www.gffarms.com/history.

39. Kraybill and Nolt, *Amish Enterprise*, 166.

40. Comaroff and Comaroff, *Ethnicity, Inc.*, 2.

41. Busch, "Moral Economy of Grades and Standards," 273.

42. Hoffmann, "U.S. Food Safety Policy," 28.

43. Aden A. Yoder, "From Inside the Office," *GFF Newsletter* 11, no. 1 (Mar. 2015): 1.

44. Right to Know is a nonprofit organization that works nationally to promote information about the impact of GMOs.

45. Open-pollinated seeds are produced by natural, random pollination by wind, birds, or insects. Hybrid seeds are produced by mechanically crossing two strong varieties to generate seed that will show vigorous growth when planted. Genetically modified plants are produced by inserting a desired portion of the DNA of an unrelated organism into the target plant's DNA in a laboratory setting.

46. Certification by the Non-GMO Project, a nonprofit that independently verifies non-GMO ingredients, does not necessarily mean that a plant was not grown using chemical fertilizers or pesticides or that an animal was not raised using artificial hormones.

47. Jimmy Doherty, "Amish Farmers Embrace GM Crops," *BBC News*, 24 Nov. 2008, accessed 15 Nov. 2017, http://news.bbc.co.uk/2/hi/science/nature/7742471.stm.

48. Doherty, "Amish Farmers Embrace GM Crops."

49. Mark Stoll, "Is There Death in the Pot?," *Family Life*, Mar. 2011, 11.

50. A 2014 overview of more than seventeen hundred original research studies on GM crops acknowledged that the debate remains intense but concluded that "the scientific research conducted thus far has not detected any significant hazards directly connected with the use of GE crops." Nicolia et al., "Overview of the Last 10 Years," 77.

51. Stone points out, however, that the health risks of DDT were not scientifically discerned until more than sixty years after its introduction. Glenn Davis Stone, "GM Foods: A Moment of Honesty," *Food, Farming and Biotechnology* (blog), 29 July 2015, https://fieldquestions.com/2015/07/29/gm-foods-a-moment-of-honesty/.

52. Brunk and Coward, *Acceptable Genes*, examines religious attitudes toward bioengineered food and diet across many subpopulations within pluralistic societies.

53. Kraybill, Johnson-Weiner, and Nolt, *Amish*, 288.

CHAPTER 5: The Forest for the Trees

1. Bumgardner, Romig, and Luppold, "Amish Furniture Cluster in Ohio," 130.

2. The number of Amish working in wood-related industries varies across settlements. Kraybill and Nolt, *Amish Enterprise*, 45, found that in Lancaster, Pennsylvania, in 1993, 34 percent of 118 enterprises surveyed related to wood products, and their list did not include Amish in timber buying, logging, sawmill operation, or lumber sales. Even so, they concluded that "woodworking trades comprise the largest clustering of enterprises." In contrast, Kanagy and Kraybill, "Rise of Entrepreneurship," 275, found that 69.5 percent of the men in Indiana County, Pennsylvania, in 1999 were engaged in wood-related businesses. Since these studies were made, the number of enterprises involving wood products has continued to climb across Amish communities.

3. Bumgardner, Romig, and Luppold, "Amish Furniture Cluster in Ohio," 133. In the Holmes County settlement in 2005, the median number of employees in shops building furniture or other items out of sawn lumber was 4.0, with a range from 1 to 105.

4. Buehlmann and Schuler, "U.S. Household Furniture Industry," 25n: "The Amish manufacturers, taking advantage of the lack of other suppliers of such 'handcrafted,' customized fur-

niture . . . offer an example of successfully competing as a U.S.-based manufacturer in a global market."

5. "The Green Machine," *Ohio's Amish Country Magazine*, 4 Aug. 2015, http://www.ohiosamishcountry.com/article/the-green-machine.html.

6. Quartersawing lumber enhances the pattern of the grain. The log is first cut lengthwise into four quarters. Then each quarter is cut radially from the bark to the center of the section.

7. Widmann, "Forests of Ohio, 2015," 2.

8. Terence E. Hanley, "Timber Stand Improvement (TSI)," http://ruralaction.org/wp-content/uploads/2012/02/FM-VIII-Timber-Stand-Improvement.pdf.

9. Heasley and Guries, "Forest Tenure and Cultural Landscapes," 190.

10. See Hershberger, *Chainsaws and Big Bucks*, for an account of techniques to enhance forests for deer, such as hinge-cutting to create bedding zones.

11. In Ohio, participants in these workshops, which cover forest safety and environmental protection, receive "master forester" certification.

12. Ellefson, Kilgore, and Granskog, "Government Regulation of Forestry Practices," 623.

13. Ellefson, Kilgore, and Granskog, "Government Regulation of Forestry Practices," 623.

14. Brian MacGowen and Duane McCoy, "Forestry Best Management Practices," makes the point that "installing BMPs on a timber sale is the right thing to do to be a good steward of the land." *Woodland Steward* 23, no. 2 (Fall 2014), http://www.inwoodlands.org/forestry-best-management-pract/.

15. As noted in Kraybill, Johnson-Weiner, and Nolt, *Amish*, 354, "The overriding narrative of Amish encounters with the state is one of principled obedience."

16. Olshan, "National Amish Steering Committee," 73.

17. Bumgardner, Romig, and Luppold, "Amish Furniture Cluster in Ohio," 77. While 5 percent of the firms surveyed by the authors of this paper had annual sales of more than $3 million, two-thirds of the reporting businesses had annual sales of $500,000 or less.

18. Bumgardner et al., "How Clustering Dynamics Influence Lumber Utilization Patterns," 79.

19. Kanagy and Kraybill, "Rise of Entrepreneurship," 275–76, notes that when agricultural opportunities are limited, the number of heads of household engaged in the wood-products industry rises.

20. Whitney, *From Coastal Wilderness to Fruited Plain*, 77.

21. Robinson, *Forest and the Trees*, 11.

22. Whitney, *From Coastal Wilderness to Fruited Plain*, 191.

23. Whitney, *From Coastal Wilderness to Fruited Plain*, 191.

24. In 1891 Congress established the National Forest system. The academic discipline of forestry appeared, and foresters in the newly minted United States Forest Service sought to protect and maintain forested habitats under governmental jurisdiction. See Robinson, *Forest and the Trees*, 13.

25. Heasley and Guries, "Forest Tenure and Cultural Landscapes," 182.

26. Rametsteiner and Simula, "Forest Certification," 87.

CHAPTER 6: Tinkering with Creation

1. The North American Deer Registry records the DNA genotypes of whitetail deer in the captive population and makes this information available to breeders. See http://deerregistry.com/.

2. Lauren Slater and Vincent J. Musi, "Wild Pets: The Debate over Owning Exotic Animals," *National Geographic Magazine*, Apr. 2014.

3. Bush, Baker, and Macdonald, "Global Trade in Exotic Pets, 2006–2012," 663–65.

4. Originally called the Mid Ohio Exotic Animal and Bird Sale, the name was changed in 2010 to Mid Ohio Alternative Animal and Bird Sale, in part to deflect criticism from animal-rights activists.

5. Photographs are prohibited on the auction grounds, so published photos are taken clandestinely.

6. The announcement for the fall 2016 sale can be found at http://www.mthopeauction.com/sale/mid-ohio-alternative-animal-and-bird-sale/2016/09/16.

7. Hurst and McConnell, *Amish Paradox*, 200.

8. Kyle Swenson, "Amish Dog Breeders Face Heat," *Cleveland Scene*, 21 July 2010, https://www.clevescene.com/cleveland/amish-dog-breeders-face-heat/Content?oid=1953690.

9. Swenson, "Amish Dog Breeders Face Heat." See also "Amish Puppy Mills Exposed," New York State Citizens Against Puppy Mills, www.citizensagainstpuppymills.org/pmamish.php.

10. Kraybill, Nolt, and Weaver-Zercher, *Amish Way*, 147.

11. Eric Sandy, "Ohio's Dog Breeding Regulations Aren't Solving the Puppy Problem," *Cleveland Scene*, 7 Oct. 2015.

12. The number of state-licensed kennels in Ohio is far smaller than would be needed to meet demand by brokers for puppies. The Ohio Department of Agriculture had estimated that the new law would apply to 800 to 1,000 breeders in the state, but by 2015 only 182 kennels had sought licenses, which apply to breeders who have nine or more breeding dogs and sell more than sixty dogs a year. Sandy, "Ohio's Dog Breeding Regulations."

13. In Lancaster County, Pennsylvania, Amish dog breeders are, in fact, sometimes confused with Old Order Mennonites and often end up carrying the media blame for all Plain groups who breed dogs.

14. Erik Wesner, "Amishman Plans Dog Business; Activists Fear Puppy Mill," *Amish America* (blog), 11 Dec. 2014, http://amishamerica.com/colorado-amishman-plans-dog-business-locals-fear-puppy-mill/.

15. Horse descriptions from *The Fourth Annual Mid-Ohio Memorial Trotting Sale*, catalog, 2016, www.mthopeauction.com/sites/default/files/2016 Memorial Trotting Catalog_0.pdf.

16. Kraybill, Johnson-Weiner, and Nolt, *Amish*, 14.

17. High-end horse breeding has not yet spread to the western Amish settlements. Although western Amish use horses not just for buggies but for managing their cattle, for riding, for back-country camping, and for hunting expeditions to the high country, there is not the strong demand for fancy buggy horses that there is in the East.

18. At the 2016 Mid-Ohio Memorial Catalogued Trotting Sale, horses sold for an average of $12,000-$15,000, and a horse named Ozzy Ozzy brought in $140,000.

19. An account of the event can be found in Mary Ann Sherman, "Horse Progress Days," *Farming*, Fall 2014, 42–43.

20. Statement on the website advertising Horse Progress Days, http://www.horseprogressdays.com/welcome.asp.

21. Advertisement in *The Fourth Annual Mid-Ohio Memorial Trotting Sale*, catalog, 2016, 9.

22. D'Arcy Egan, "Holmes County Deer Hunting Preserve Ordered to Euthanize Herd of 300 Trophy Bucks, Does," 5 Dec. 2014, http://www.cleveland.com/outdoors/index.ssf/2014/12/holmes_county_deer_hunting_pre.html.

23. Egan, "Holmes County Deer Hunting Preserve."

24. Art Holden, "CWD Positives Now Up to 19: Division of Wildlife Looking to Take 3-Mile Radius Samples," *Wooster (OH) Daily Record*, 27 July 2015.

25. Egan, "Holmes County Deer Hunting Preserve."

26. Ohio Department of Natural Resources, 3 Oct. 2016, http://wildlife.ohiodnr.gov/wildlife-home/post/odnr-continues-plan-to-monitor-ohio-s-deer-herd-for-chronic-wasting-disease.

27. "White-tailed Deer Production," PennState Extension, http://extension.psu.edu/business/ag-alternatives/livestock/exotic-livestock/white-tailed-deer-production; Matt Reese, "Deer Farms Thriving in Ohio," *Ohio Country Journal*, 11 June 2014.

28. Ryan Sabalow, "Buck Fever," *Indiana Star*, 27 Mar. 2014.
29. Sabalow, "Buck Fever."
30. Sabalow, "Buck Fever."
31. Sabalow, "Buck Fever."
32. Amish deer farmers breed and sell so-called shooter bucks to private preserves. But only a few Amish also own or share ownership of the exclusive high-fence hunting preserves, where clients may pay tens of thousands of dollars for a guided deer hunt.
33. The Boone and Crockett Club website shows how the score for a deer antler rack is calculated. http://www.boone-crockett.org/bgRecords/ScoringYourTrophy.asp?area=bgRecords&ID=416327E9&se=1.
34. The website for various whitetail auctions is http://auctions.whitetailsales.com/upcoming-auctions.html.
35. Ryan Sabalow, "Is the Rack Worth the Risk?," *Indiana Star*, 27 Mar. 2014.
36. Largely because of CWD, the national Quality Deer Management Association, with chapters in many Amish communities, has publicly opposed deer farming, leading some Amish deer farmers to leave the organization. See Art Holden, "Caught in the Middle," *Wooster (OH) Daily Record*, 9 Apr. 2012.
37. In Ohio, more than half the members listed on the Whitetail Deer Farmers of Ohio website had addresses in the Holmes County Amish settlement, a clear indication that Amish in the industry identify political lobbying as important to their business interests. See the Whitetail Deer Farmers of Ohio website, accessed 14 Dec. 2016, http://www.wdfo.com.
38. Abe Miller, "Hello to All Fellow Dog Breeders," *Ohio Professional Dog Breeders Association Newsletter* 6, no. 4 (Oct. 2015): 7.
39. "Benefits of Being a OPDBA Member," *Ohio Professional Dog Breeders Association Newsletter* 6, no. 4 (Oct. 2015): 25.
40. In his classic 1859 book *On the Origin of Species*, Charles Darwin repeatedly used the production of new variants in animal husbandry as evidence to advance his arguments about the process of speciation.
41. The emerging Old Order churches of the Midwest issued a statement condemning the use of mules by the Lancaster churches: "We declare it to be unseemly for one who professes to be a Christian to mingle God's creatures, such as the horse and the donkey, which produces the mule, since God did not create such in the beginning." See "Agreement of Thirty-Four Ministers," 259.

CHAPTER 7: Bringing Nature Home

1. Bhatti and Church, "Cultivating Natures," 366.
2. Bhatti and Church, "Cultivating Natures," 367.
3. Hondagneu-Sotelo, "Cultivating Questions for a Sociology of Gardens," 511.
4. Beachy, *Unser Leit*, 1:146.
5. Genesis 2:9.
6. Amos 4:9.
7. The full text reads: "The kiss of the sun for pardon / the song of the birds for mirth / one is nearer God's heart in a garden / than anywhere else on earth." These are the last lines of a poem called "God's Garden," written in 1913 by Dorothy Frances Gurney, a British hymn writer and poet.
8. Robbins and Sharp, "Lawn-Chemical Economy and Its Discontents," 160.
9. Paul Yoder, "Ortgeards (Orchards)," *Pilgrim's Pathway*, May–June 2015, 11 and 25.
10. Regina F. Graham, "Amish Girl, 12, Is 'Cancer Free' Two Years after Her Family Was Forced into Hiding over Court Battle to Force Her to Have Chemotherapy, as Judge Drops Case," *Daily Mail* (London), 9 Oct. 2015.
11. Kraybill, Johnson-Weiner, and Nolt, *Amish*, 339–45, describes these three resources in some

detail but also includes a fourth, the church community, which pays visits, sends circle letters, and anoints with oils.

12. Kraybill, Johnson-Weiner, and Nolt, *Amish*, 343.

13. Hurst and McConnell, *Amish Paradox*, 256.

14. See Quillin, *Wisdom of Amish Folk Medicine*, for a nostalgic view of Amish herbal remedies as a repository of knowledge from a simpler past.

15. Stein et al., "Innate Immunity and Asthma Risk," found that a sample of Amish farm children in Indiana had markedly lower rates of allergies and asthma compared with Hutterite children living on industrialized farms. These results show that Amish children are "among the least allergic subgroup ever measured in the developed world." See Moises Velasquez-Manoff, "Health Secrets of the Amish," *New York Times*, 3 Aug. 2016, https://www.nytimes.com/2016/08/04/opinion/health-secrets-of-the-amish.html.

16. We see many parallels with the use of natural medicines in urban Ecuador as described in Miles, "Science, Nature, and Tradition."

17. Davis-Floyd, *Birth as an American Rite of Passage*, 100.

18. Kraybill, Johnson-Weiner, and Nolt, *Amish*, 336.

19. Rachel Weaver is not Amish but promotes home-based health remedies that can be concocted from available plants and foodstuffs.

20. Many other herbal remedies from the Geauga settlement, including onions sprinkled with sugar for a hacking cough and jewelweed for itching, are described in Byler, *Plain and Happy Living*.

21. For example, the study reported in Kolacz et al., "Effects of Burns and Wounds (B&W)," followed five Amish burn patients who used a combination of B&W Ointment and a burdock-leaf dressing and found that dressing changes caused minimal or no pain, none of the burns became infected, and healing took less than fourteen days. The authors recommend B&W Ointment as an acceptable alternative to conventional care for first- and second-degree burns.

22. The Swartzentruber Amish were the exception, with 90 percent of respondents indicating they collected wild plants for herbal remedies.

23. Kriebel, *Powwowing among the Pennsylvania Dutch*, 16.

24. Kriebel, *Powwowing among the Pennsylvania Dutch*, 16.

25. Kraybill, Johnson-Weiner, and Nolt, *Amish*, 340, reports that fifteen self-taught Amish folk healers were identified in the liberal northern Indiana Amish enclave in the 1980s.

26. Reed et al., "Vitamin and Supplement Use." Herbal-supplement use was associated with lower use of medications, but the use of vitamin and mineral supplements was not.

27. The field of epigenetics is a new and active area of genetic research based on the idea that changes in an organism can be caused by modifications in gene expression rather than in the underlying genetic code itself. The idea that changes in lifestyle, such as diet, can address these problems and restore the body to a more natural state is very appealing to Amish (and non-Amish) folk healers. In reality, the molecular mechanisms that control gene expression, such as methylation, are extremely complex, and the chemical tags that turn genes on and off are still poorly understood. See Heard and Martienssen, "Transgenerational Epigenetic Inheritance," for an analysis of common myths surrounding epigenetics.

28. *Connection*, July 2013, inside cover. The price of a bottle of one hundred capsules was $29.95 for #1 and $29.95 for #2. Like many ads, this one appropriates scientific language, referencing "the famous Dr. Harvey Kellogg, who once said, 'Of the 22,000 operations I have performed, I have never found a single normal colon.' " Alongside a phone number and address for ordering are the line "Gott segne Dich!" (God Bless You) and a horse-and-buggy silhouette with the phrase "Serving among the Amish for 15 years."

29. For a thoughtful overview of both the dangers and the potential benefits of the doctrine

of signatures, see Matt Simon, "Fantastically Wrong: The Strange History of Using Organ-Shaped Plants to Treat Disease," *Wired*, 16 July 2014.

30. See Arvind Suresh, "Don't Believe the Biohype: Tips for Evaluating Claims about Genetics and Biotech," *Health News Review*, 26 Mar. 2016.

31. Nixon, *Slow Violence*.

32. Callahan, *False Hopes*, 135.

CHAPTER 8: Fin, Fur, and Feather

1. Christine L. Pratt, "73 Arrested for Underage Consumption: Amish Party Raided by Holmes Law Enforcement," *Wooster (OH) Daily Record*, 6 Sept. 2016.

2. Stevick's *Growing Up Amish* points out that rumspringa does not automatically equate to "wild living," that most Amish parents do not condone rumspringa, and that it does not end when one joins the church, but rather upon marriage.

3. Art Holden, "America's Best Bowstrings: Mullett Turns His $1000 Start-up Business into a Product Known around the World," *Wooster (OH) Daily Record*, 15 Oct. 2012.

4. One Old Order bishop told us the story of Jacob Hostetler's family, which gets "preached a lot." The family was attacked by Indians, and his wife and daughter were killed and his boys taken captive after he refused to allow them to shoot back.

5. An Old Order man underscored the strong belief in the right of gun ownership for hunting, estimating that "20 percent [of the Amish] are members" of the National Rifle Association.

6. Tharp, "Valued Amish Possessions," 42.

7. The Swartzentruber Amish, however, will not wear orange camo and hunt mostly on family farms without all the high-tech preparation.

8. Boglioli, *Matter of Life and Death*, 78.

9. Boglioli, *Matter of Life and Death*, 79.

10. Wengerd, "Joy of Taking the Next Generation Hunting," 165.

11. Looking for sheds means walking the woods to search for antlers that bucks have shed because of a drop in testosterone levels after the rut or after the mating season.

12. Robert Troyer, "More Than Meat for the Family," *Plain Interest* 8, no. 11 (Nov. 2008): 1.

13. Art Holden, "Walleye and Charity: A Winning Combo," *Wooster (OH) Daily Record*, 16 June 2014.

14. Art Holden, "Still in the Business: 'Fur Al' Continues His Longstanding Tradition of Hosting Baltic-Area Trapper Education Workshop," *Wooster (OH)Daily Record*, 11 Sept. 2012.

15. According to an Amish bishop in a Kentucky settlement, men from three church districts created a contest with a fifteen-dollar entry fee to see who could kill the most coyotes. Over a three-year period the contestants collectively killed three hundred animals, much to the concern of local wildlife officials, who viewed their efforts as upsetting the balance of nature.

16. Biolsi, "Imagined Geographies," 244, notes that the policy of checkerboarding was based on the colonialist assumption that white settlers would serve as role models in the project of "civilizing the Indians" and that eventually native peoples and their lands would be assimilated.

17. With respect to the first charge, Amish fishing expeditions weren't necessarily keeping more fish than allowed. For many years, the limit was six trout per person per day for fishing in a mountain lake on reservation land. "So if you take four boys, that's thirty fish a day," reflected one Amish leader. "We kind of came to the conclusion as a group that it's one of the greatest things in the world to be able to do that, and we don't want to deplete the lakes. We want it to be there for our grandchildren."

18. B. I. Azure, "Salish Pend d'Oreille Elders Air Differences with Amish Elders," *Char-Koosta News*, 20 Oct. 2011.

19. We followed up on this point with several tribal members. One man who agreed that it was a big concern said, "Our kids don't get dirty like they used to. There's too much social media, a hundred TV channels."

20. Henry Troyer Jr., "A Black Rail in Coshocton County, Ohio," *Journey of Wings*, Fall 2014, 12.

21. Jim McCormac, "Lark Bunting, Take 2," *Ohio Birds and Biodiversity* (blog), 12 Nov. 2012, http://jimmccormac.blogspot.com/search?q=Amish.

22. Stambaugh, "Rare Birders Find Rare Birds."

23. In 2009 one of the pioneers of the bicycle big year, James E. Yoder, of Walnut Creek, Ohio, put 3,900 "birding miles" on his bike and spent 785 hours in search of birds. He ended up with 252 species, just behind his birding buddy Kevin Kline, who found 254. James E. Yoder, "Birding Big Year," *Bobolink*, Winter 2009/2010, 13–16.

24. J. D. Shrock, "The Saddle-Up Cowgirls at Charm Days," Amish Leben, http://amishleben.com/2013/10/the-saddle-up-cowgirls-at-charm-days/.

25. In 2001, Kraybill, *Riddle of Amish Culture*, 70, noted that horseback riding was "generally discouraged because it borders on a worldly form of sport."

26. Serpell, *In the Company of Animals*, 168.

27. Milton, *Loving Nature*.

CHAPTER 9: Observing and Writing Nature

1. In a survey of Old Order Amish in the Holmes County settlement, Hurst and McConnell, *Amish Paradox*, 136, found that one-third of respondents had traveled to more than twenty-one states in their lifetime, while two-thirds had traveled to more than eight states.

2. The anthropologist John Hostetler argued that because the Amish live in a "high context" culture, literacy is de-emphasized. See Hostetler, *Amish Society*, 18.

3. The anthropologist Alma Gottlieb, in "Americans' Vacations," argues that tourists seek to invert class hierarchies and everyday work roles in one of two ways. Upper-class individuals want to become a "Peasant for a Day," while working-class individuals desire to be a "Queen/King for a Day" (173).

4. According to MacCannell, *Tourist*, 3, "Reality and authenticity are thought to be elsewhere: in other historical periods and in other cultures, in purer, simpler lifestyles."

5. See Meyers, "Amish Tourism," for an alternative view in which economic motivations were paramount among visitors to Shipshewana, Indiana.

6. Amish individuals differ, of course, in their reaction to this visibility. Some say they enjoy being stopped and asked if they're Amish, but others do not like being objects of curiosity. Many Amish privately poke fun at tourists who gawk at them.

7. Gingerich, *History of Pinecraft*.

8. Kraybill, Johnson-Weiner, and Nolt, *Amish*, 23.

9. Trollinger and Trollinger, *Righting America at the Creation Museum*, offers an analysis of how Answers in Genesis, a conservative evangelical think tank, takes "real science" and reframes it so that it is consistent with a young-earth perspective. They note that an Amish construction crew was involved in building the museum's latest attraction, a life-size Noah's Ark.

10. Many examples also exist of an extended family or other groups chartering their own bus and, if necessary, advertising for additional passengers.

11. Rebecca Miller, "An Amish Girl's Adventures: Mackinac Island," *AmishAmerica* (blog), 10 Nov. 2015, http://amishamerica.com/an-amish-girls-adventures-mackinac-island/.

12. Aaron Miller, "Arizona Birding Trip," *Bobolink*, Summer 2009, 11–12.

13. Edwin Troyer, "Screaming Elk," *Hometown Outdoors* 5, no. 2 (Mar.–Apr. 2016): 11–13.

14. Lyng, "Edgework and the Risk-Taking Experience."

15. Stoddart and Nezhadhossein, "Is Nature-Oriented Tourism a Pro-Environmental Practice?" 544, cautions against assuming that nature-oriented tourism inevitably aligns with environmentalism "both as a social movement and as a personal worldview." Haluza-DeLay, "Nothing Here to Care About," found a diminished motivation to care for their home environment after young people spent time in the wilderness because they came to view cities as "already wrecked" (47).

16. See Kraybill, Johnson-Weiner, and Nolt, *Amish*, 375, for a sample list of these publications.

17. Johnson-Weiner, "Publish or Perish," 201.

18. Kraybill, Johnson-Weiner, and Nolt, *Amish*, 372.

19. In addition to Pathway, other major publishers of Amish works include Gordonville Print House in Lancaster County, Pennsylvania; Study Time in the Elkhart-LaGrange settlement in Indiana; and Carlisle Printing in Walnut Creek, Ohio.

20. Hochstetler and Hochstetler, *Life on the Edge of the Wilderness*, 3.

21. Joe Bontrager and Rosemary Bontrager, "From the Bird Sanctuary," *Connection*, Mar. 2014, 58.

22. Jamin Schrock, "Sketch a Bird," *Journey of Wings*, Fall 2016, 8.

23. Kraybill, Johnson-Weiner, and Nolt, *Amish*, 373.

24. "This is My Father's World," *Shining for Jesus* 2, no. 4 (Apr. 2016): 20.

25. "Are You an Invertebrate?," *Family Life*, Feb. 2010, 19. The morality tale in this example is clearly more important than an accurate scientific definition of an invertebrate, which is actually a cold-blooded animal without a backbone.

26. Quotations are from *Pencil Passion*, a brochure advertising Andy Mast's work.

27. For an account of Mast's personal and professional journey, see Virginia Daffron, "Passion from his Pencil: Amish Fine Artist Andy Mast," *Virginia Daffron* (blog), 15 Aug. 2016, http://www.virginiadaffron.com/features/passion-from-his-pencil-amish-fine-artist-andy-mast/.

28. Amy Schlabach, "His World," *Farming*, Fall 2016, 52.

29. Kraybill, Nolt, and Weaver-Zercher, *Amish Way*, 139.

30. Jim Van Der Pol, "Conversations with the Land," *Farming*, Summer 2013, 10.

31. David Kline, "Editor's Note," *Farming*, Fall 2013, 6.

32. McKibben, introduction, xxii–xxiii.

33. Berry, *Devoted to Nature*.

34. Kline, *Scratching the Woodchuck*, 85.

35. Kline, *Scratching the Woodchuck*, 180.

36. David Kline, "Letter from Larksong," *Farming*, Fall 2009, 7.

37. Kline, *Scratching the Woodchuck*, 14.

38. Kline, *Scratching the Woodchuck*, 88.

39. Kline, *Scratching the Woodchuck*, 162.

40. Kline, *Great Possessions*, 219.

41. David Kline, "Letter from Larksong," *Farming*, Summer 2011, 7.

42. David Kline, "Letter from Larksong," *Farming*, Winter 2001, 4.

43. David Kline, "Letter from Larksong," *Farming*, Fall 2004, 7.

44. David Kline, "Letter from Larksong," *Farming*, Fall 2010, 7.

45. David Kline, "Letter from Larksong," *Farming*, Spring 2012, 7.

46. Burke, Welch-Devine, and Gustafson, "Nature Talk in an Appalachian Newspaper," concluded from a study of environmental discourse in print media in Appalachia that "the environment is generally represented as an amenity to be enjoyed rather than a subject of concern, that environmental degradation, when represented at all, is often discussed in vague or distancing terms, and that human agency is typically presented in individualizing, hyper-local terms rather than in collective, community- or national-scale ones" (185).

CHAPTER 10: Acting Locally

1. Hardin, "Tragedy of the Commons," 1244, gives the example of herdsmen who share a common pasture. When each herdsman tries to maximize his gain by adding animals to his flock, the result is depletion of the pasture. Thus, "freedom in a commons brings ruin to all." See Berkes et al., "Benefits of the Commons," 72–73, for a critique of Hardin's thesis.

2. Kristin Bialik, "Most Americans Favor Stricter Environmental Laws and Regulations," Pew Research Center, 14 Dec. 2016, http://www.pewresearch.org/fact-tank/2016/12/14/most-americans-favor-stricter-environmental-laws-and-regulations/.

3. See Kraybill, Johnson-Weiner, and Nolt, *Amish*, 352–68, for a detailed analysis of Amish relations with the government, including relations with the national steering committee.

4. Kraybill, Johnson-Wiener, and Nolt, *Amish*, 353.

5. Kraybill, Johnson-Wiener, and Nolt, *Amish*, 353–54.

6. Kraybill, "Negotiating with Caesar," 15.

7. Place, "Land Use," 194.

8. See Environmental Protection Agency, "Impaired Waters and TDMLs," https://www.epa.gov/tmdl/program-overview-total-maximum-daily-loads-tmdl#1.

9. See "Little Elkhart River Watershed Management Plan Addendum," 29; and http://www.chesapeakebay.net/track/restoration.

10. Mercury comes mostly from atmospheric depositions, while PCBs, banned since 1979, are residues from industrial sites where electronics were produced. Thus, herbicides are the primary chemical contaminant associated with Amish farms. See "Toxic Contaminants in the Chesapeake Bay and its Watershed," http://executiveorder.chesapeakebay.net/ChesBayToxics_finaldraft_11513b.pdf.

11. Nonpoint pollution comes from the landscape instead of from a single polluting source. See Environmental Protection Agency, "Nonpoint Source: Agriculture," https://www.epa.gov/nps/nonpoint-source-agriculture.

12. Weaver, Moore, and Parker, "Farmer Learning Circle," 210.

13. See Reagan Waskom and David J. Cooper, "Why Farmers and Ranchers Think the EPA Clean Water Rules Go Too Far," *Conversation*, 27 Feb. 2017.

14. See Weaver, Moore, and Parker, "Farmer Learning Circle," 208–9, for a similar account of non-Amish farmers' reaction to the EPA in Ohio.

15. In *Problems of Plenty*, J. Douglas Hurt describes how over the course of the twentieth century American farmers lost their autonomy because federal aid always came with conditions.

16. "Little Elkhart River Watershed Management Plan Addendum," 5.

17. Environmental Protection Agency, "Reducing Livestock-Induced Pollution in Emma Creek," https://www.epa.gov/sites/production/files/2015-11/documents/in-emma.pdf.

18. "Little Elkhart River Watershed Management Plan Addendum," 5.

19. Meyers and Nolt, *Amish Patchwork*, 141, acknowledges the new connections between Amish and the state that have emerged in northern Indiana in recent years.

20. Environmental Protection Agency, "Reducing Livestock-Induced Pollution in Emma Creek."

21. The 2002 report stated: "Observed aquatic resource degradation from agriculture included direct manure and urine discharge to streams, milking waste discharged by pipe to streams, dumping of fruit processing waste into streams, direct habitat alteration by dredging and cattle walking in streams, and lack of wooded riparian corridor." State of Ohio Environmental Protection Agency, "Total Maximum Daily Loads for the Sugar Creek Basin: Final Report."

22. The grants covered water-quality sampling, educational outreach, and collaborative learn-

ing and problem-solving with local farmers in the watershed. "The Sugar Creek Method," http://sugarcreekmethod.osu.edu/to8_pageview3/Sugar_Creek_Method.htm.

23. US congressman Bob Gibbs, of the Seventh District, which includes Holmes County, has been an outspoken critic of the burdensome regulations created by the Clean Water Act. See Paul Quinlan, "Anti-Environmental House Freshman Leads Charge against Obama's Clean Water Agenda," *New York Times*, 3 May 2011.

24. They were skeptical, for example, about the claim that agriculture was responsible for 70 percent of the watershed's pollution. See Weaver, Moore, and Parker, "Farmer Learning Circle."

25. Widner, *Old Order Amish Beliefs*, 58. Swartzentruber farmers in this study reported that they let their cows walk in the stream because the cows needed clean water and that they put manure on their fields every two to three weeks. But they said that they fished and swam in the river and no people or cattle had gotten sick.

26. Comito, Wolseth, and Morton, "State's Role in Water Quality," argue that Soil and Water Conservation Districts are largely a conservative force that maintains the agricultural status quo.

27. The fact that the money for remediation came from the cheese company and not the government also made a difference in how the program was received. "How Alpine Cheese Used Water Quality Trading to Achieve Clean Water Goals," *Cheese Reporter*, 1 May 2014, 22, http://npaper-wehaa.com/cheese-reporter/2014/05/?g=print#?article=2216537.

28. Amanda Peterka, "Amish Farmers in Chesapeake Bay Watershed Find Themselves in EPA's Sights," *New York Times*, 10 Oct. 2011.

29. Rona Kobell, "The Amish: Makers of Fine Jam, Cabinetry, and Polluted Rivers," *Grist*, 6 Nov. 2014.

30. See "Watson Run Summary Paper," https://www.epa.gov/sites/production/files/documents/watson_run_summary_paper.pdf. The results are also summarized in Sindya N. Bhanoo, "Amish Farms Draw Rare Government Scrutiny," *New York Times*, 8 June 2010.

31. Daniel Walmer, "Amish Targeted over Environmental Problems," *Washington Times*, 21 Jan. 2017, http://www.washingtontimes.com/news/2017/jan/21/amish-targeted-over-environmental-problems/.

32. In fact, according to Hostetler, *Amish Society*, 142, as early as 1990 a survey of wells near Amish farms found that 67 percent had unsafe nitrate levels.

33. John Luciew, "The Amish and the Chesapeake: A Plain Farmer Adopts Modern Methods," Pennlive.com, 3 Jan. 2013, http://www.pennlive.com/midstate/index.ssf/2013/01/the_amish_and_the_chesapeake_1.html.

34. In 2012, Pennsylvania tobacco production was valued at more than $43 million. See National Agricultural Statistics Service, "Pennsylvania Crop Production," USDA news release, 2012.

35. Chesapeake Bay Foundation, *State of the Bay, 2016*, http://www.cbf.org/about-the-bay/state-of-the-bay-report-2016.

36. Chesapeake Bay Foundation, "CBF Prioritizes Five Pennsylvania Counties to Jumpstart Cleanup Effort," 13 Sept. 2016, http://www.cbf.org/news-media/newsroom/pa/2016/09/13/cbf-prioritizes-five-pennsylvania-counties-to-jumpstart-cleanup-efforts.

37. Kobell, "Amish."

38. Molly Redden, "The Amish Are Getting Fracked: Their Religion Prohibits Lawsuits—and the Energy Companies Know It," *New Republic*, 6 June 2013.

39. See John Funk, "Rural Ohio Is the Wild West as Gas and Oil Companies Compete for Drilling Rights," *Plain Dealer*, 27 Oct. 2011.

40. Pennsylvania alone has nearly eight thousand active wells. http://stateimpact.npr.org/pennsylvania/drilling/.

41. Julie Carr Smyth and Kevin Begos, "Tradition and Temptation as the Amish Debate Fracking," *Boston Herald*, 9 July 2013.

42. See Hudgins, "Fracking's Future in a Coal Mining Past," for an account of how community members with deep ties to the coal industry in southwestern Pennsylvania describe the diffuse and transient workers associated with fracking as moving in like "aliens."

43. Humes, *Garbology*, 5, 6.

44. Kidder, "Role of Outsiders," 216–17.

45. Megan Durisin, "13 Money Secrets from The Amish," *Business Insider*, 5 May 2014, http://www.businessinsider.com/money-secrets-of-amish-people-2014-4.

46. Effie Mullet, Fredericksburg, OH, in *Budget*, 27 Apr. 2016, 32.

47. Berry, *Devoted to Nature*, 186.

48. Ulrich-Schad, Brock, and Prokopy, "Comparison of Awareness, Attitudes, and Usage."

CHAPTER 11: Thinking Globally

1. For a summary of the issues that global environmentalists prioritize, see Ripple et al., "World Scientists' Warning to Humanity."

2. Place, "Land Use," 191.

3. Hostetler, *Amish Society*, 142.

4. In Middlefield Township, Ohio, Amish business leaders circulated a ballot petition in 2015 for a measure to loosen restrictions on the size of home-based businesses. Though the ballot measure failed, a compromise was reached.

5. Of the 640 million federally owned acres, the Bureau of Land Management administers 41 percent; the Forest Service, 30 percent; and the Fish and Wildlife Service and the National Park Service, 28 percent. See Vincent, Hanson, and Argueta, *Federal Land Ownership*, 4.

6. Vincent, Hanson, and Argueta, *Federal Land Ownership*, 21.

7. The Holmes County (Ohio) Rails to Trails runs from Fredericksburg to Millersburg and is being extended to the south. The Indiana bat (*Myotis sodalis*), an endangered species, roosts under the bark of trees such as shagbark hickory (*Carya ovata*) and white oak (*Quercus alba*).

8. Wilson, "Little Things."

9. The northern spotted owl requires western old-growth forests. Invoking the Endangered Species Act, the Forest Service in 1990 limited logging, provoking a heated controversy about preservation and use. For a short summary, see Claire Andre and Manuel Velasquez, "The Spotted Owl Controversy," 13 Nov. 2015, https://www.scu.edu/ethics/focus-areas/more/environmental-ethics/resources/ethics-and-the-environment-the-spotted-owl/.

10. Ellwood et al., "Early Flowering"; McEwan et al., "Flowering Phenology Change."

11. Parmesan and Yohe, "Globally Coherent Fingerprint"; Scheffers et al., "Broad Footprint of Climate Change."

12. For an analysis of the link between climate change and extreme weather, see Diffenbaugh et al., "Quantifying the Influence of Global Warming."

13. The global growth rate of 1.09 percent (for 2018) is from the Worldodometers website, accessed 5 May 2018, http://www.worldometers.info/world-population/world-population-by-year/.

14. Donnermeyer, Anderson, and Cooksey, "Amish Population," 75.

15. Donnermeyer, Anderson, and Cooksey, "Amish Population," 90.

16. As of 2017, the newest Amish settlements are in Maine and in Texas. See the map of Amish settlements in chapter 1.

17. The residents meant by this that the growing season was longer than in nearby areas.

18. Vitousek et al., "Human Domination of Earth's Ecosystems," provides considerable evidence that human population growth is presently a major driver of environmental degradation.

Murtaugh and Schlax, "Reproduction," notes that current reproduction massively multiplies the future impact of humans on the earth.

19. Miller et al., "Health Status, Health Conditions," found that of 249 women who responded to questions about birth control, 20.9 percent of them were using birth control. About half of these women and their partners were using condoms.

20. Campanella, Korbin, and Acheson, "Pregnancy and Childbirth among the Amish."

21. Morton et al., "Pediatric Medicine."

22. Dunlap et al., "Measuring Endorsement of the New Ecological Paradigm"; Hawcroft and Milfont, "Use (and Abuse) of the New Environmental Paradigm Scale."

23. The NEP survey was a part of our household-resource-use survey. If Amish respondents agreed to be interviewed after they took our survey, we then had the opportunity to query them about the logic behind their responses.

24. Hawcroft and Milfont, "Use (and Abuse) of the New Environmental Paradigm Scale," presents a meta-analysis of 69 published papers, assessing 139 different populations, in the United States and worldwide. However, some studies used twelve or sometimes only six questions. Hawcroft and Milfont found that shorter versions of the fifteen-question survey systematically produced lower NEP scores. To counteract this possible bias, for comparative purposes we used only those twelve published studies that surveyed populations in the United States and that used the fifteen-question version of the survey.

25. Hawcroft and Milfont, "Use (and Abuse) of the New Environmental Paradigm Scale." The sixty-nine studies reviewed by these authors included studies from the United States and many other countries.

26. Schultz, Unipan, and Gamba, "Acculturation and Ecological Worldview," used only twelve survey questions, so their results are not strictly comparable to ours.

27. Amburgey and Thoman, "Dimensionality of the New Ecological Paradigm."

28. For all but concept 3, English and Amish mean scores were significantly different (p = 0.000). For concept 3, the scores were not significantly different.

29. See the appendix for a discussion of scale reliability among our Amish and English respondents.

30. However, for question 2, asking whether humans have the right to modify the environment to meet their needs, Amish and English scores were statistically no different, with scores around 3.0. In our postsurvey interviews, Amish respondents wanted to ask what counted as modifying the environment. Their attitude was that if it meant clearing a woodlot to make a farm field, they could do that, but if it meant constructing a massive dam to flood a huge area, it shouldn't be done.

31. Hawcroft and Milfont, "Use (and Abuse) of the New Environmental Paradigm Scale." For this comparison, we include studies in the United States and in other countries. Of fifty-eight comparable samples (using the fifteen-question survey) from around the world, only six had mean NEP scores lower than that of the English population in our study.

CHAPTER 12: Parochial Stewards

1. The hope that the Amish inspire among some onlookers—that even a white, settler society in a North American context can achieve a degree of ecological sustainability and harmony with nature—may also be linked, paradoxically, to the particulars of Amish ethnicity, history, and geography. We are grateful to Steve Nolt for calling attention to this dimension of nostalgia for the Amish.

2. Conklin and Graham argue in "Shifting Middle Ground" that the alliance between international environmental organizations and Amazonian Indians in Brazil has rested on the problematic assumption that "native peoples' views of nature and ways of using natural resources are consistent with Western conservationist principles" (696).

3. Barry Commoner, an American ecologist who died in 2012, is credited with articulating the four rules of ecology. In his book *The Closing Circle*, 33–48, he lists them as (1) everything is connected to everything else; (2) everything must go somewhere; (3) nature knows best; and (4) there is no such thing as a free lunch.

4. A correlation between religious conservatism and antienvironmental attitudes has been amply demonstrated. Arbuckle and Konisky, "Role of Religion in Environmental Attitudes," found that reading the Bible literally was inversely related to expression of environmental concern, while Hand and Van Liere, "Religion, Mastery-over-Nature, and Environmental Concern," concluded that those affiliated with religiously conservative traditions (e.g., Baptists and Mormons) were more likely to endorse the idea of a God-given mastery over nature than were individuals involved in liberal denominations (e.g., Episcopalians and Methodists).

5. Gillespie, "How Culture Constructs Our Sense of Neighborhood," 23. The sketches were drawn by fifteen Amish and fifteen non-Amish children in the same area in an Old Order Pennsylvania settlement. The sketch maps of non-Amish pupils "demonstrated a less restricted perception of neighborhood, often veering away from the immediate home and family to depict parks, neighbors' houses, and even whole communities."

6. See Kalmann, "Environmental Stewardship," 319–22.

7. Quoted in Tik Root, "An Evangelical Movement Takes on Climate Change," *Newsweek*, 9 Mar. 2016.

8. Haenn, "Power of Environmental Knowledge," 336–37.

9. Jones and Dunlap, "Social Bases of Environmental Concern," found that the social positions of those who supported environmental protection were remarkably stable from 1970 to 1990. On average, those who were more supportive were younger, more educated, politically liberal, and Democrats, had been raised and were living in urban areas, and were not employed in industries focused on resource extraction.

10. Naylor, "Pinchot, Gifford (1865–1946)," 1280.

11. Merchant, "Gender and Environmental History."

12. Naylor, "Pinchot, Gifford (1865–1946)," 1281.

13. Steve Grant, "Gifford Pinchot: Bridging Two Eras of National Conservation," ConnecticutHistory.org, https://connecticuthistory.org/gifford-pinchot-bridging-two-eras-of-national-conservation/.

14. According to some scholars, including Grim and Tucker, in *Ecology and Religion*, and Gottlieb, in *Greener Faith*, a significant resacralization of nature has occurred in Christianity in recent years. However, Konisky, "Greening of Christianity?," finds little support for increased concern about the environment by Christians in longitudinal data from Gallup's annual surveys. Even so, many religious organizations of diverse theological orientations are actively promoting stewardship of the environment. Sponsel, *Spiritual Ecology*, describes a quiet revolution that encompasses "a breadth and diversity of religions and spiritualities" (xiii), while Taylor, *Dark Green Religion*, charts the rise of broadly defined "religious" communities that consider "nature to be sacred, imbued with intrinsic value, and worthy of reverent care" (ix).

15. Kalland, "Religious Environmentalist Paradigm," 1368.

16. TEK focuses on the practices of indigenous people that help to manage resources with minimal environmental harm and that seem to demonstrate a prescientific but accurate understanding of ecological processes.

17. Orlove and Brush, "Anthropology and the Conservation of Biodiversity," 335.

18. Kalland, "Religious Environmentalist Paradigm," 1370. As one example, Kalland points out that if humans and spirits have a close relationship, exploitation of nature can be accomplished by enticing spirits to move to other locations or appeasing them.

19. Smith and Wishnie, "Conservation and Subsistence in Small-Scale Societies," 509.

20. Kennedy et al., "Why We Don't 'Walk the Talk,'" 151.

21. Hunn, "Mobility as a Factor."

22. Pope Francis, *Laudato si'*, para. 21.

23. Pope Francis, *Laudato si'*, paras. 33, 140.

24. *Green Bible*. The foreword is by Desmond Tutu, and contributors include a diverse group of clergy, scholars, and environmental leaders.

25. Peter Steinfels, "Evangelical Group Defends Laws Protecting Endangered Species as a Modern 'Noah's Ark,'" *New York Times*, 31 Jan. 1996.

26. Caldicott, "Conservation Values and Ethics," 120.

27. Caldicott, "Conservation Values and Ethics," 119.

28. Caldicott, "Conservation Values and Ethics."

29. One reason why many Amish resist a more inclusive definition of stewardship is that they associate it with threatening aspects of evangelism. Some evangelical churches actively try to recruit Amish to join their ranks, and almost all allow technologies and organizational structures forbidden by the Amish.

30. Olshan, "Homespun Bureaucracy."

31. Agrawal, *Environmentality*, showed that in India the state, ostensibly harnessed to improve life for the nation's citizens, exerted control in capricious ways that often disenfranchised the poor. Similar processes, such as the placing of contaminated waste sites in poor neighborhoods, have been well documented in the United States.

32. Kraybill, Johnson-Weiner, and Nolt, *Amish*, 354.

33. As early as 1956, Huntington, "The Amish Farmer and the Government Agricultural Worker," 27, noted that farmers and government workers had little in common and that their ability to work together was dependent on the attitudes of both sides. More recently, Brock, Ulrich-Schad, and Prokopy, "Bridging the Divide," documented continued challenges in the relation between the Amish and extension workers, as well as some surprising success stories.

34. Welk-Joerger, "Religion, Science, and Diverse Farming Methods."

35. Giddens, *Consequences of Modernity*, 8.

36. See McNeill, *Environmental History*, for a compelling analysis of the environmental changes humans have wrought on the earth over the past one hundred years.

37. I=PAT was proposed by Paul Ehrlich and John Holdren in conversation with Barry Commoner in the early 1970s. It was hailed as a breakthrough because it conceived of environmental degradation more broadly than as just pollution and as caused by multiple forces. In the decades since, it has been criticized as relying on a notion of carrying capacity that may not have the same relevance for human populations that it does for other organisms. For example, studies have shown that growing population and affluence can be partially offset by technological efficiencies. See Chertow, "IPAT Equation and Its Variants," 13.

38. Murtaugh and Schlax, "Reproduction and the Carbon Legacies," 14.

39. Moledina et al., "Amish Economic Transformations."

40. Kraybill, Johnson-Weiner, and Nolt, *Amish*, 410.

41. Ritzer, *McDonaldization of Society*, 141.

42. Berry, *Our Only World*, 62.

43. Paulson, "Degrowth," 425. The degrowth movement is an eclectic array of research projects and social movements that seek "an equitable downscaling of production and consumption that increases human well-being and enhances ecological conditions." Schneider, Kallis, and Martinez-Alier, "Crisis or Opportunity?," 511.

44. In combatting algal blooms, for example, there is already strong evidence that regimes of accountability make a difference. Chesapeake Bay remediation efforts, which have teeth in them, have been much more effective than voluntary restrictions in the Great Lakes.

45. Weaver-Zercher, *Amish in the American Imagination*, 144.

46. Kalland, "Religious Environmentalist Paradigm," 1370.

47. Berry, *Hidden Wound*.

48. Kidder, "Role of Outsiders," argues that outsiders routinely misunderstand Amish simplicity as religious asceticism rather than the product of consensus about the effects a given technology would have on family and community life. Similarly, Johnson-Weiner, "Old Order Amish Education," argues that even the 1972 Supreme Court decision in *Wisconsin v. Yoder* was based on an outdated view of the Amish as farmers completely isolated from the surrounding world.

49. Carrier, "Protecting the Environment the Natural Way," 401.

50. Pedersen, "Nature, Religion, and Cultural Identity," 272.

51. Denying the Noble Other doesn't make the Amish the Ignoble Other. As Richard White points out in *Middle Ground*, x, "Diverse peoples adjust their differences through what amounts to creative, and often expedient, misunderstandings. . . . They often misinterpret and distort both the values and practices of those they deal with, but from these misunderstandings arise new meanings and from them new practices—the shared meanings and practices of the middle ground."

APPENDIX: Methods

1. IBM SPSS Statistics for Macintosh, version 23.0 and 25.0, released 2014 and 2016, respectively.

2. For natural gas, U.S. Energy Information Administration, https://www.eia.gov/dnav/ng/ng_pri_rescom_dcu_SOH_a.htm; for electric power, U.S. Energy Information Administration, https://www.eia.gov/electricity/monthly/epm_table_grapher.cfm?t=epmt_5_6_a; for gasoline, GasBuddy, http://www.ohiogasprices.com/retail_price_chart.aspx.

3. Dr. Jessica Bedore, personal conversation, 28 Jan. 2017. We also had input from several Amish friends.

4. Camargo, Ryan, and Richard, "Energy Use and Greenhouse Gas Emissions," 266. The calculator can be downloaded at Farm Energy Analysis Tool (FEAT), www.ecologicalmodels.psu.edu/agroecology/feat/download.htm.

5. Kristensen et al., "Carbon Footprint of Cheese," 233.

6. Dong et al., "Emissions from Livestock and Manure Management."

7. For emission factors of fuels, Environmental Protection Agency Center for Corporate Climate Leadership, https://www.epa.gov/sites/production/files/2016-09/documents/emission-factors_nov_2015_v2.pdf.

8. The calorie counter used to estimate servings from calories was at WebMD, http://www.webmd.com/diet/healthtool-food-calorie-counter.

9. When Amish parents pass the family property to one of their children, the elders build a smaller grandparents' house (*Dawdi Haus* in Pennsylvania Dutch) attached to or adjacent to the homestead, creating a multigenerational family living unit.

10. Dunlap et al., "Measuring Endorsement of the New Ecological Paradigm," 426; Hawcroft and Milfont, "Use (and Abuse) of the New Ecological Paradigm Scale," 144; Johnson, Bowker and Cordell, "Ethnic Variation in Environmental Belief and Behavior," 167.

11. Amburgey and Thoman, "Dimensionality of the New Ecological Paradigm," 245.

Agrawal, Arun. *Environmentality: Technologies of Government and the Making of Subjects.* Durham, NC: Duke University Press, 2005.

"Agreement of Thirty-Four Ministers, Holmes County, Ohio, June 1, 1865." In *Proceedings of the Amish Ministers' Meetings, 1862–1878*, ed. Paton Yoder and Steven R. Estes, 258–60. Goshen, IN: Mennonite Historical Society, 1999.

Amburgey, Jonathan W., and Dustin B. Thoman. "Dimensionality of the New Ecological Paradigm: Issues of Factor Structure and Measurement." *Environment and Behavior* 44 (2012): 235–56.

"Amish Population Change, 1992–2017." Amish Studies. https://groups.etown.edu/amishstudies.

Anderson, Cory. "Amish Education: A Synthesis." *Journal of Amish and Plain Anabaptist Studies* 3, no. 1 (2015): 1–24.

Arbuckle, M. B., and D. M. Konisky. "The Role of Religion in Environmental Attitudes." *Social Science Quarterly* 96 (2015): 1244–63.

Ayres, Robert U. "Life Cycle Analysis: A Critique." *Resources, Conservation and Recycling* 14 (1995): 199–223.

Bailey, Connor, Leif Jensen, and Elizabeth Ransom. "Rural America in a Globalizing World: Introduction and Overview." In *Rural America in a Globalizing World: Problems and Prospects for the 2010s*, ed. Bailey, Jensen, and Ransom, xiii–xxix. Morgantown: West Virginia University Press, 2014.

Baugh, Amanda J. *God and the Green Divide: Religious Environmentalism in Black and White.* Berkeley: University of California Press, 2016.

Beachy, Leroy. *Unser Leit: The Story of the Amish.* 2 vols. Millersburg, OH: Goodly Heritage Books, 2011.

Beck, Ulrich. "Ecological Questions in a Framework of Manufactured Uncertainties." In *The New Social Theory Reader: Contemporary Debates*, ed. Steven Seidman and Jeffrey C. Alexander, 267–75. New York: Routledge, 2001.

Bender, Sue. *Plain and Simple: A Woman's Journey to the Amish.* San Francisco: HarperCollins, 1989.

Berkes, Fikret, David Feeney, Bonnie J. McCay, and James M. Acheson. "The Benefits of the Commons." In Haenn, Wilk, and Harnish, *Environment in Anthropology*, 68–74.

Berry, Evan. *Devoted to Nature: The Religious Roots of American Environmentalism.* Berkeley: University of California Press, 2015.

Berry, Thomas. *The Dream of the Earth.* San Francisco: Sierra Club Books, 1988.

Berry, Wendell. *Bringing It to the Table: On Farming and Food.* Berkeley, CA: Counterpoint, 2009.

______. *The Hidden Wound.* Berkeley, CA: Counterpoint, 2010.

______. *Our Only World: Ten Essays.* Berkeley, CA: Counterpoint, 2015.

Bhatti, Mark, and Andrew Church. "Cultivating Natures: Homes and Gardens in Late Modernity." *Sociology* 35 (2001): 365–83.

Biersack, Aletta. "Reimagining Political Ecology: Culture/Power/History/Nature." In *Reimagining Political Ecology*, ed. Biersack and James B. Greenberg, 3–40. Durham, NC: Duke University Press, 2006.

Billig, Michael S., and Elam Zook. "The Functionalist Problem in Kraybill's *Riddle of Amish Culture*." *Journal of Amish and Plain Anabaptist Studies* 5, no. 1 (2017): 82–95.

Biolsi, Thomas. "Imagined Geographies: Sovereignty, Indigenous Space, and American Indian Struggles." *American Ethnologist* 32, no. 2 (2005): 239–59.

Blake, Katharine V., Enrico A. Cardamone, Steven D. Hall, Glenn R. Harris, and Susan M. Moore. "Modern Amish Farming as Ecological Agriculture." *Society and Natural Resources* 10 (1997): 143–59.

Boglioli, Marc. *A Matter of Life and Death: Hunting in Contemporary Vermont.* Amherst: University of Massachusetts Press, 2009.

Bourne, Joel K., Jr. *The End of Plenty: The Race to Feed a Crowded World*. New York: Norton, 2016.
Brock, Caroline, Jessica D. Ulrich-Schad, and Linda Prokopy. "Bridging the Divide: Challenges and Opportunities for Public Sector Agricultural Professionals Working with Amish and Mennonite Producers on Conservation. *Environmental Management* 61, no. 5 (2018): 756–71.
Brunk, Conrad G., and Harold Coward, eds. *Acceptable Genes: Religious Traditions and Genetically Modified Foods*. Buffalo: SUNY Press, 2009.
Buehlmann, Urs, and Al Schuler. "The U.S. Household Furniture Industry: Status and Opportunities." *Forest Products Journal* 59, no. 9 (2009): 20–28.
Bumgardner, Matthew S., Gary W. Graham, P. Charles Goebel, and Robert L. Romig. "How Clustering Dynamics Influence Lumber Utilization Patterns in the Amish-Based Furniture Industry in Ohio." *Journal of Forestry* 109, no. 2 (Mar. 2011): 74–81.
Bumgardner, Matthew S., Robert L. Romig, and William Luppold. "The Amish Furniture Cluster in Ohio: Competitive Factors and Wood Use Estimates." In *Proceedings of the 16th Central Hardwoods Forest Conference*, ed. Douglass F. Jacobs and Charles H. Michler, GTR-NRS-P-24 (Newtown Square, PA: USDA Forest Service, Northern Research Station, 2008), 130–38.
______. "Wood Use by Ohio's Amish Furniture Cluster." *Forest Products Journal* 57, no. 12 (2007): 6–12.
Burke, Brian J., Meredith Welch-Devine, and Seth Gustafson. "Nature Talk in an Appalachian Newspaper: What Environmental Discourse Analysis Reveals about Efforts to Address Exurbanization and Climate Change." *Human Organization* 74, no. 1 (2015): 185–96.
Busch, Lawrence. "The Moral Economy of Grades and Standards." *Journal of Rural Studies* 16 (2000): 273–83.
Bush, Emma R., Sandra E. Baker, and David W. Macdonald. "Global Trade in Exotic Pets, 2006–2012." *Conservation Biology* 28, no. 3 (2014): 663–76.
Byler, Emma. *Plain and Happy Living: Recipes and Remedies*. Burton, OH: Log Cabin, 1995.
Caldicott, J. Baird. "Conservation Values and Ethics." In *Principles of Conservation Biology*, ed. Martha J. Groom, Gary K. Meffe, and C. Ronald Carroll, 111–35. 3rd ed. Sumberland, MA: Sinauer Associates, 2006.
Callahan, Daniel. *False Hopes: Overcoming the Obstacles to a Sustainable, Affordable Medicine*. New Brunswick, NJ: Rutgers University Press, 1999.
Camargo, Gustavo G. T., Matthew R. Ryan, and Tom L. Richard. "Energy Use and Greenhouse Gas Emissions from Crop Production Using the Farm Energy Analysis Tool." *BioScience* 63 (2013): 263–73.
Campanella, Karla, Jill F. Korbin, and Louise Acheson. "Pregnancy and Childbirth among the Amish." *Social Science and Medicine* 36, no. 3 (1993): 333–42.
Carlisle, Liz. "Audits and Agrarianism: The Moral Economy of an Alternative Food Network." *Elementa: Science of the Anthropocene* 3 (2015): 66, doi: 10.12952/journal.elementa.000066.
Carrier, James G. "Protecting the Environment the Natural Way: Ethical Consumption and Commodity Fetishism." In Haenn, Wilk, and Harnish, *Environment in Anthropology*, 401–10.
Cepek, Michael L. "Foucault in the Forest: Questioning Environmentality in Amazonia." *American Ethnologist* 38, no. 3 (2011): 501–15.
Chertow, Marian R. "The IPAT Equation and Its Variants." *Journal of Industrial Ecology* 4, no. 4 (Oct. 2000): 13–29.
Ching, Barbara, and Gerald W. Creed. "Recognizing Rusticity: Identity and the Power of Place." In *Knowing Your Place: Rural Identity and Cultural Hierarchy*, ed. Ching and Creed, 1–39. New York: Routledge, 1997.
Clements, Rhonda. "An Investigation of the Status of Outdoor Play." *Contemporary Issues in Early Childhood* 5, no. 1 (2004): 68–80.
Comaroff, John L., and Jean Comaroff. *Ethnicity, Inc.* Chicago: University of Chicago Press, 2009.

Comito, Jacqueline, Jon Wolseth, and Lois Morton. "The State's Role in Water Quality: Soil and Water Conservation District Commissioners and the Agricultural Status Quo." *Human Organization* 72, no. 1 (Spring 2013): 44–54.

Commoner, Barry. *The Closing Circle: Man, Nature, and Technology*. New York: Random House, 1971.

Conkin, Paul K. *A Revolution Down on the Farm*. Lexington: University Press of Kentucky, 2008.

Conklin, Beth A., and Laura R. Graham. "The Shifting Middle Ground: Amazonian Indians and Eco-Politics." *American Anthropologist* 97, no. 4 (Dec. 1995): 695–710.

Conklin, Harold. "The Study of Shifting Cultivation." *Current Anthropology* 2 (1961): 27–61.

Courtney, S. P., J. A. Blakesley, R. E. Bigley, M. L. Cody, J. P. Dumbacher, R. C. Fleischer, A. B. Franklin, J. F. Franklin, R. J. Gutiérrez, J. M. Marzluff, and L. Sztukowski. *Scientific Evaluation of the Status of the Northern Spotted Owl*. Portland, OR: Sustainable Ecosystems Institute, 2004. https://www.fws.gov/oregonfwo/species/Data/NorthernSpottedOwl/BarredOwl/Documents/CourtneyEtAl2004.pdf.

Craumer, Peter R. "Farm Productivity and Energy Efficiency in Amish and Modern Dairy Farming." *Agriculture and Environment* 4 (1979): 281–99.

Cronon, William. "The Trouble with Wilderness: Or, Getting Back to the Wrong Nature." *Environmental History* 1, no. 1 (Jan. 1996): 7–28.

Cross, John A. "The Expanding Role of the Amish in America's Dairy Industry." *FOCUS on Geography*, 50, no. 3 (2007): 7–16.

Crumley, Carole L. "Historical Ecology: A Multidisciplinary Ecological Orientation." In *Historical Ecology: Cultural Knowledge and Changing Landscapes*, ed. Crumley, 1–16. Santa Fe, NM: School of American Research, 1994.

Crutzen, Paul J. "Geology of Mankind: The Anthropocene." *Nature* 415 (2002): 23.

Davis-Floyd, Robbie E. *Birth as an American Rite of Passage*. Berkeley: University of California Press, 2004.

Demaria, Federico, Francois Schneider, Filka Sekulova, and Joan Martinez-Alier. "What is Degrowth? From an Activist Slogan to a Social Movement." In Haenn, Wilk, and Harnish, *Environment in Anthropology*, 390–400.

DeWalt, Mark W. *Amish Education in the United States and Canada*. Oxford: Rowman & Littlefield, 2006.

Diffenbaugh, Noah S., Deepti Singh, Justin S. Mankin, Daniel E. Horton, Daniel L. Swain, Danielle Touma, Allison Charland, Yunjie Liu, Matz Haugen, Michael Tsiang, and Bala Rajaratnam. "Quantifying the Influence of Global Warming on Unprecedented Extreme Climate Events." *Proceedings of the National Academy of Sciences* 114, no. 19 (9 May 2017): 4881–86.

Dong, Hongmin, Joe Mangino, Tim A. McAllister, Jerry L. Hatfield, Donald E. Johnson, Keith R. Lassey, Magda Aparecida de Lima, Anna Romanovskaya, Deborah Bartram, Daryl Gibb, and John H. Martin Jr. "Emissions from Livestock and Manure Management." In *2006 IPCC (Intergovernmental Panel on Climate Change) Guidelines for National Greenhouse Gas Inventories*, vol. 4, *Agriculture, Forestry, and Other Land Use*, ed. H. Simon Eggleston, Leandro Buendia, Kyoko Miwa, Todd Hgara, and Kiyoto Tanabe. http://www.ipcc-nggip.iges.or.jp/public/2006gl/vol4.html.

Donnermeyer, Joseph F., Cory Anderson, and Elizabeth C. Cooksey. "The Amish Population: County Estimates and Settlement Patterns." *Journal of Amish and Plain Anabaptist Studies* 1, no. 1 (Apr. 2013): 72–109.

Dove, Michael R., and Carol Carpenter. "Introduction: Major Historical Currents in Environmental Anthropology." In *Environmental Anthropology: A Historical Reader*, ed. Dove and Carpenter, 1–85. Oxford: Blackwell, 2008.

Downing, Crystal. "Witnessing the Amish: Plain People on Fancy Film." In *The Amish and the*

Media, ed. Diane Zimmerman Umble and David L. Weaver-Zercher, 25–41. Baltimore: Johns Hopkins University Press, 2008.

Drummond, Mark A., and Thomas R. Loveland. "Land-Use Pressure and a Transition to Forest-Cover Loss in the Eastern United States." *BioScience* 60 (2010): 286–98.

Dunlap, Riley E., Kent D. Van Liere, Angela G. Mertig, and Robert Emmet Jones. "Measuring Endorsement of the New Ecological Paradigm: A Revised NEP scale." *Journal of Social Issues* 56 (2000): 425–42.

Ediger, Marlow. 1998. "Teaching Science in the Old Order Amish School." *Journal of Instructional Psychology* 25, no. 1 (1998): 62–66.

Ellefson, Paul C., Michael A. Kilgore, and James E. Granskog. "Government Regulation of Forestry Practices on Private Forest Land in the United States: An Assessment of State Government Responsibilities and Program Performance." *Forest Policy and Economics* 9 (2007): 620–32.

Ellwood, Elizabeth R., Stanley A. Temple, Richard B. Primack, Nina L. Bradley, and Charles C. Davis. "Record-Breaking Early Flowering in the Eastern United States." *PLOS/ONE* 8 (2013): 1–9.

Galindo, Rene, and Constance Brown. "Narrative in an Amish Farmer's Appropriation of Nature Writing." *Written Communication* 12, no. 2 (Apr. 1995): 147–85.

Geertz, Clifford. *The Interpretation of Cultures*. New York: Basic Books, 1973.

Giddens, Anthony. *The Consequences of Modernity*. Stanford, CA: Stanford University Press, 1991.

Gillespie, Carol Ann. "How Culture Constructs Our Sense of Neighborhood: Mental Maps and Children's Perception of Place." *Journal of Geography* 109 (2010): 18–29.

Gingerich, Noah. *The History of Pinecraft, 1925–1960: A Historical Album of the Amish and Mennonites in Pinecraft, Florida*. Walnut Creek, OH: Carlisle, 2006.

Gottlieb, Alma. "Americans' Vacations." *Annals of Tourism Research* 9 (1982): 165–87.

Gottlieb, Roger. *A Greener Faith: Religious Environmentalism and Our Planet's Future*. Oxford: Oxford University Press, 2009.

The Green Bible. New rev. standard version. New York: Harper One, 2006.

Grim, John, and Mary Evelyn Tucker. *Ecology and Religion*. Washington, DC: Island Press, 2014.

Gupta, Akhil. *Red Tape: Bureaucracy, Structural Violence, and Poverty in India*. Durham, NC: Duke University Press, 2012.

Haenn, Nora. "The Power of Environmental Knowledge: Ethnoecology and Environmental Conflicts in Mexican Conservation." In Haenn, Wilk, and Harnish, *Environment in Anthropology*, 332–43.

Haenn, Nora, Richard R. Wilk, and Allison Harnish, eds. *The Environment in Anthropology: A Reader in Ecology, Culture, and Sustainable Living*. 2nd ed. New York: New York University Press, 2016.

Haluza-DeLay, Randolph. "Nothing Here to Care About: Participant Constructions of Nature Following a 12-Day Wilderness Program." *Journal of Environmental Education* 32, no. 4 (2001): 43–48.

Hand, Carl M., and Kent D. Van Liere. "Religion, Mastery-over-Nature, and Environmental Concern." *Social Forces* 63 (1984): 555–70.

Hardin, Garrett. "The Tragedy of the Commons." *Science* 162 (1968): 1243–48.

Hawcroft, L. J., and T. L. Milfont. "The Use (and Abuse) of the New Environmental Paradigm Scale over the Past 30 Years: A Meta-Analysis." *Journal of Environmental Psychology* 30 (2010): 143–58.

Heard, Edith, and Robert Martienssen. "Transgenerational Epigenetic Inheritance: Myths and Mechanisms." *Cell* 157, no. 1 (2014): 95–109. doi: http://dx.doi.org/10.1016/j.cell.2014.02.045.

Heasley, Lynne, and Raymond P. Guries. "Forest Tenure and Cultural Landscapes: Environmental Histories in the Kickapoo Valley." In *Social Conflict over Property Rights: Who Owns America?*, ed. Harvey M. Jacobs, 182–207. Madison: University of Wisconsin Press, 1998.

Hershberger, Henry. *Chainsaws and Big Bucks: Creating Trophy Properties.* Fredericksburg, OH: Holmes, 2010.

Hochstetler, Martin, and Susan Hochstetler. *Cabin Life on the Kootenai.* N.p.: privately printed, 1991.

______. *Life on the Edge of the Wilderness.* N.p.: privately printed, 1987.

Hoffmann, Sandra. "U.S. Food Safety Policy Enters a New Era." *Amber Waves* 9, no. 4 (2011): 24–39.

Hondagneu-Sotelo, Pierrette. "Cultivating Questions for a Sociology of Gardens." *Journal of Contemporary Ethnography* 39 (2010): 498–516.

Honey, Martha. "Treading Lightly? Ecotourism's Impact on the Environment." In Haenn, Wilk, and Harnish, *Environment in Anthropology*, 380–89.

Hostetler, John A. *Amish Society.* 4th ed. Baltimore: Johns Hopkins University Press, 1993. First published in 1963.

Hostetler, John A., and Gertrude Enders Huntington. *Amish Children: Education in the Family, School and Community.* 2nd ed. New York: Harcourt, 1992.

Hudgins, Anastasia. "Fracking's Future in a Coal Mining Past: Subjectivity Undermined." *Culture, Agriculture, Food, and Environment* 35, no. 1 (June 2013): 54–59.

Hughes, J. Donald. "Europe as Consumer of Exotic Biodiversity: Greek and Roman Times." *Journal of Landscape Research* 28, no. 1 (2003): 21–31.

Humes, Edward. *Garbology: Our Dirty Love Affair with Trash.* New York: Penguin, 2012.

Hunn, Eugene S. "Mobility as a Factor Limiting Resource Use in the Columbian Plateau of North America." In *Resource Managers: North American and Australian Foragers*, ed. Nancy M. Williams and Eugene S. Hunn, 17–43. London: Routledge, 1981.

Huntington, Gertrude Enders. "The Amish Farmer and the Government Agricultural Worker." *Yale Conservation Studies* 5 (1956): 27–29.

Hurst, Charles E., and David L. McConnell. *An Amish Paradox: Diversity and Change in the World's Largest Amish Community.* Baltimore: Johns Hopkins University Press, 2010.

Hurt, R. Douglas. *Problems of Plenty: The American Farmer in the Twentieth Century.* Chicago: Ivan R. Dee, 2002.

Imhoff, Daniel, and Jo Ann Baumgartner. Introduction to *Farming and the Fate of Wild Nature: Essays in Conservation-Based Agriculture*, ed. Imhoff and Baumgartner, v–viii. Healdsburg, CA: Watershed Media, 2006.

Institute in Basic Life Principles. *Character Sketches, Vol. 1.* Oak Brook, IL: IBLP, 1976.

Jackson, Mary. "Amish Agriculture and No-Till: The Hazards of Applying USLE to Unusual Farms." *Journal of Soil and Water Conservation* 43, no. 6 (1988): 483–86.

Jackson, Wes. *New Roots for Agriculture.* Berkeley, CA: North Point, 1980.

Johnson, Cassandra Y., J. M. Bowker, and H. Ken Cordell. "Ethnic Variation in Environmental Belief and Behavior: An Examination of the New Ecological Paradigm in a Social Psychological Context." *Environment and Behavior* 36 (2004): 157–86.

Johnson-Weiner, Karen M. *New York Amish: Life in the Plain Communities of the Empire State.* Ithaca, NY: Cornell University Press, 2010.

______. "Old Order Amish Education: The *Yoder* Decision in the 21st Century." *Journal of Amish and Plain Anabaptist Studies* 3, no. 1 (2015): 25–44.

______. "Publish or Perish: Amish Publishing and Old Order Identity." In *The Amish and the Media*, ed. Diane Zimmerman Umble and David L. Weaver-Zercher, 201–19. Baltimore: Johns Hopkins University Press, 2008.

______. "Technological Diversity and Cultural Change among Contemporary Amish Groups." *Mennonite Quarterly Review* 88 (Jan. 2014): 5–22.

______. *Train Up a Child: Old Order Amish Education.* Baltimore: Johns Hopkins University Press, 2006.

Jones, Robert Emmet, and Riley E. Dunlap. "The Social Bases of Environmental Concern: Have They Changed Over Time?" *Rural Sociology* 57, no. 1 (1992): 28–47.

Kalland, Arne. "Environmentalism and Images of the Other." In *Nature Across Cultures: Views of Nature and the Environment in Non-Western Cultures*, ed. Helaine Selin, 1–17. New York: Springer, 2003.

______. "The Religious Environmental Paradigm." In *Encyclopedia of Religion and Nature*, ed. Bron Taylor, 1367–71. New York: Continuum, 2005.

Kalmann, Rowenn Beth. "Environmental Stewardship and the Production of Subjectivities: Indigenous, Scientific, and Economic Rationalities in Ancash, Peru." PhD diss., Michigan State University, 2017.

Kanagy, Conrad L., and Donald B. Kraybill. "The Rise of Entrepreneurship in Two Old Order Communities." *Mennonite Quarterly Review* 70, no. 3 (1996): 263–79.

Keim, John W. *Comfort for the Burned and Wounded.* Quakertown, PA: Philosophical Publishing, 1999.

Kennedy, Emily Huddart, Thomas M. Beckley, Bonita L. McFarlane, and Solange Nadeau. "Why We Don't 'Walk the Talk': Understanding the Environmental Values/Behaviour Gap in Canada." *Human Ecology Review* 16, no. 2 (Winter 2009): 151–60.

Kidder, Robert L. "The Role of Outsiders." In Kraybill, *Amish and the State*, 213–33.

Kimball, Abigail R., Andrea Schuhmann, and Hutch Brown. "More Kids in the Woods: Reconnecting Americans with Nature." *Journal of Forestry* 107, no. 7 (Oct./Nov. 2009): 373–77.

Kingsolver, Barbara. *Animal, Vegetable, Miracle: A Year of Food Life*. New York: Harper Perennial, 2008.

Kline, David. *Great Possessions: An Amish Farmer's Journal.* San Francisco: North Point, 1990.

______. *Scratching the Woodchuck: Nature on an Amish Farm.* Athens: University of Georgia Press, 1997.

Kolacz, N. M., M. T. Jaroch, M. L. Bear, and R. F. Hess. "The Effects of Burns and Wounds (B&W) / Burdock Leaf Therapy on Burn-Injured Amish Patients: A Pilot Study Measuring Pain Levels, Infection Rates, and Healing Times." *Journal of Holistic Nursing* 32, no. 4 (Dec. 2014): 327–40. doi: 10.1177/0898010114525683.

Kolbert, Elizabeth. *The Sixth Extinction: An Unnatural History*. New York: Henry Holt, 2014.

Konisky, David M. "The Greening of Christianity? A Study of Environmental Attitudes over Time." *Environmental Politics* 27, no. 2 (2018): 267–91.

Kraybill, Donald B., ed. *The Amish and the State*. 2nd ed. Baltimore: Johns Hopkins University Press, 2003.

______. "Amish Informants: Mediating Humility and Publicity." In *The Amish and the Media*, ed. Diane Zimmerman Umble and David L. Weaver-Zercher, 161–78. Baltimore: Johns Hopkins University Press, 2008.

______. "Negotiating with Caesar." In Kraybill, *Amish and the State*, 3–20.

______. "Plotting Social Change across Four Affiliations." In Kraybill and Olshan, *Amish Struggle with Modernity*, 53–74.

______. *The Riddle of Amish Culture*. Rev. ed. Baltimore: Johns Hopkins University Press, 2001.

Kraybill, Donald B., Karen M. Johnson-Weiner, and Steven M. Nolt. *The Amish*. Baltimore: Johns Hopkins University Press, 2014.

Kraybill, Donald B., and Steven M. Nolt. *Amish Enterprise: From Plows to Profits*. 2nd ed. Baltimore: Johns Hopkins University Press, 2004.

Kraybill, Donald B., Steven M. Nolt, and David L. Weaver-Zercher. *The Amish Way: Quiet Faith in a Perilous World*. San Francisco: Jossey-Bass, 2010.

Kraybill, Donald B., and Marc A. Olshan, eds. *The Amish Struggle with Modernity*. Hanover, NH: University Press of New England, 1994.

Krech, Shepard, III. *The Ecological Indian: Myth and History*. New York: Norton, 1999.

Kriebel, David W. *Powwowing among the Pennsylvania Dutch: A Traditional Medical Practice in the Modern World*. State College: Pennsylvania State University Press, 2016.

Kristensen, Troels, Karen Søegaard, Jørgen Eriksen, and Lisbeth Mogensen. "Carbon Footprint of Cheese Produced on Milk from Holstein and Jersey Cows Fed Hay Differing in Herb Content." *Journal of Cleaner Production* 101 (2015): 229–37.

LeFebvre, Henri. *The Production of Space*. Trans. Donald Nicholson-Smith. Oxford: Blackwell, 1991.

Little, Paul E. "Environments and Environmentalisms in Anthropological Research: Facing a New Millennium." *Annual Review of Anthropology* 28 (1999): 253–84.

"Little Elkhart River Watershed Management Plan Addendum." Prepared by David Arrington, LaGrange County Soil and Water Conservation District. Aug. 2009. https://www.in.gov/idem/nps/files/wmp_littleelkhart_7-182_addendum.pdf.

Loewen, Royden. "The Quiet on the Land: The Environment in Mennonite Historiography." *Journal of Mennonite Studies* 23 (2005): 151–64.

Logsdon, Gene. "Did the Amish Get It Right After All?" *Contrary Farmer*, 13 Jan. 2009.

______. *Living at Nature's Pace: Farming and the American Dream*. White River Junction, VT: Chelsea Green, 2000.

Lopez, Barry. *The Rediscovery of North America*. New York: Random House, 1990.

Louv, Richard. *Last Child in the Woods: Saving Our Children from Nature-Deficit Disorder*. Rev. ed. Chapel Hill, NC: Algonquin Books, 2008.

Lowrey, Sean, and Allen G. Noble. "The Changing Occupational Structure of the Amish of the Holmes County, Ohio, Settlement." *Great Lakes Geographer* 7, no. 1 (2000): 25–37.

Luke, Timothy W. "On Environmentality: Geo-Power and Eco-Knowledge in the Discourse of Contemporary Environmentalism." In *The Environment in Anthropology: A Reader in Ecology, Culture, and Sustainable Living*, ed. Nora Haenn and Richard R. Wilk, 257–69. New York: New York University Press, 2006.

Luthy, David. *The Amish in America: Settlements That Failed, 1840–1960*. Aylmer, ON: Pathway, 1986.

______. *Why Some Amish Communities Fail: Extinct Settlements, 1961–1999*. Rev. ed. Aylmer, ON: Pathway, 2000.

Lyng, Stephen. "Edgework and the Risk-Taking Experience." In *Edgework: The Sociology of Risk-Taking*, ed. Lyng, 3–16. New York: Routledge, 2005.

MacCannell, Dean. *The Tourist: A New Theory of the Leisure Class*. New York: Schocken, 1976.

MacLeish, Archibald. "If Life Means Going Without, the Amish Go Without." In *Amish Roots: A Treasury of History, Wisdom, and Lore*, ed. John A. Hostetler, 280–82. Baltimore: Johns Hopkins University Press, 1989.

Mariola, Matthew, and David McConnell. "The Shifting Landscape of Amish Agriculture: Balancing Tradition and Modernity in an Amish Organic Produce Cooperative." *Human Organization* 72, no. 2 (2013): 144–53.

Marx, Leo. *The Machine in the Garden: Technology and the Pastoral Ideal in America*. 35th anniv. ed. New York: Oxford University Press, 2000.

McConnell, David L., and Charles E. Hurst. "No 'Rip Van Winkles' Here: Amish Education since Wisconsin v Yoder." *Anthropology and Education Quarterly* 37, no. 3 (2006): 236–54.

McEwan, Ryan W., Robert J. Brecha, Donald R. Geiger, and Grace P. John. "Flowering Phenology Change and Climate Warming in Southwestern Ohio." *Plant Ecology* 212, no. 1 (2011): 55–61.

McKibben, Bill. *Deep Economy*. New York: Henry Holt, 2007.

______. *The End of Nature*. New York: Random House, 1989.

______. Introduction to *American Earth: Environmental Writing since Thoreau*, ed. McKibben, xxi–xxxi. New York: Library of America, 2008.

McNeill, J. R. *An Environmental History of the Twentieth-Century World*. New York: Norton, 2000.
Merchant, Carolyn. "Gender and Environmental History." *Journal of American History* 76 (Mar. 1990): 1117–21.
Meyers, Thomas J. "Amish Tourism: 'Visiting Shipshewana is Better Than Going to the Mall.'" *Mennonite Quarterly Review* 77 (Jan. 2003): 109–26.
______. "Lunch Pails and Factories." In Kraybill and Olshan, *Amish Struggle with Modernity*, 165–81.
Meyers, Thomas J., and Steven M. Nolt. *An Amish Patchwork: Indiana's Old Orders in the Modern World*. Bloomington: Indiana University Press, 2005.
Michener, Charles D. *The Social Behavior of the Bees*. Cambridge, MA: Belknap Press of Harvard University Press, 1974.
Miles, Ann. "Science, Nature, and Tradition: The Mass Marketing of Natural Medicine in Urban Ecuador." *Medical Anthropology Quarterly* 12, no. 2 (1998): 206–25.
Miller, Daniel. *At Home in Hickory Hollow*. Winesburg, OH: Legacy, 2010.
Miller, Kirk, Berwood Yost, Sean Flaherty, Marianne M. Hillemeier, Gary A. Chase, Carol S. Weisman, and Anne-Marie Dyer. "Health Status, Health Conditions, and Health Behaviors among Amish Women: Results from the Central Pennsylvania Women's Health Study (CePAWHS)." *Women's Health Issues* 17 (2007): 162–71.
Milton, Kay. "Cultural Theory and Environmentalism." In Haenn, Wilk, and Harnish, *Environment in Anthropology*, 250–53.
______. *Environmentalism and Cultural Theory: Exploring the Role of Anthropology in Environmental Discourse*. New York: Routledge, 1996.
______. *Loving Nature: Towards an Ecology of Emotion*. New York: Routledge, 2002.
Moledina, Amyaz A., David L. McConnell, Stephanie A. Sugars, and Bailey R. Connor. "Amish Economic Transformations: New Forms of Income and Wealth Distribution in a Traditionally 'Flat' Community." *Journal of Amish and Plain Anabaptist Studies* 2, no. 1 (2014): 1–22.
Montgomery, David R. *The Rocks Don't Lie: A Geologist Investigates Noah's Flood*. New York: Norton, 2012.
Moore, Richard. "Sustainability and the Amish: Chasing Butterflies?" *Culture, Agriculture, Food, and Environment* 16, no. 53 (Dec. 1995): 24–25.
Morton, D. Holmes, Caroline S. Morton, Kevin A. Strauss, Donna L. Robison, Erik G. Puffenberger, Christine Hendrickson, and Richard I. Kelley. "Pediatric Medicine and the Genetic Disorders of the Amish and Mennonite People of Pennsylvania." *American Journal of Medical Genetics* 121C, no. 1 (2003): 5–17.
Muir, John. *The Yosemite*. New York: Century, 1912.
Murtaugh, Paul A., and Michael G. Schlax. "Reproduction and the Carbon Legacies of Individuals." *Global Environmental Change* 19 (2009): 14–20.
Nash, Roderick. *Wilderness and the American Mind*. 4th ed. New Haven, CT: Yale University Press, 2001.
Naylor, D. Keith. "Pinchot, Gifford (1865–1946)." In *Encyclopedia of Religion and Nature*, ed. Bron Taylor, 1280–81. New York: Continuum, 2005.
Nazarea, Virginia D., ed. *Ethnoecology: Situated Knowledge/Located Lives*. Tucson: University of Arizona Press, 1999.
Netting, Robert McC. *Smallholders, Householders: Farm Families and the Ecology of Intensive, Sustainable Agriculture*. Stanford, CA: Stanford University Press, 1993.
Nicolia, Alessandro, Alberto Manzo, Fabio Veronesi, and Daniele Rosellini. "An Overview of the Last 10 Years of Genetically Engineered Crop Safety Research." *Critical Reviews in Biotechnology* 34, no. 1 (2014): 77–88.
Nixon, Rob. *Slow Violence and the Environmentalism of the Poor*. Cambridge, MA: Harvard University Press, 2013.

Nolt, Steven M. *A History of the Amish*. Rev. ed. Intercourse, PA: Good Books, 2003.
______. "Who Are the Real Amish? Rethinking Diversity and Identity among a Separate People." *Mennonite Quarterly Review* 82 (July 2008): 377–94.
Olshan, Marc A. "Homespun Bureaucracy: A Case Study in Organizational Evolution." In Kraybill and Olshan, *Amish Struggle with Modernity*, 199–213.
______. "Modernity, the Folk Society, and the Old Order Amish." In Kraybill and Olshan, *Amish Struggle with Modernity*, 185–96.
______. "The National Amish Steering Committee." In Kraybill, *Amish and the State*, 67–86.
______. "What Good Are the Amish?" In Kraybill and Olshan, *Amish Struggle with Modernity*, 231–42.
Olshan, Marc A., and Kimberly D. Schmidt. "Amish Women and the Feminist Conundrum." In Kraybill and Olshan, *Amish Struggle with Modernity*, 215–29.
Orlove, Benjamin S., and Stephen B. Brush. "Anthropology and the Conservation of Biodiversity." *Annual Review of Anthropology* 25 (Oct. 1996): 329–52.
Orr, David W. "Political Economy and the Ecology of Childhood." In *Children and Nature: Psychological, Sociocultural, and Evolutionary Investigations*, ed. Peter H. Kahn and Stephen R. Kellert, 279–303. Cambridge, MA: MIT Press, 2002.
Palmgren, Claire R., M. Granger Morgan, Wandi Bruine de Bruin, and David W. Keith. "Initial Public Perceptions of Deep Geological and Oceanic Disposal of Carbon Dioxide." *Environmental Science and Technology* 38 (2004): 6441–50.
Parajuli, Pramod. "How Can Four Trees Make a Jungle?" *Terra Nova* 3, no. 3 (1989): 15–31.
Parmesan, Camille, and G. Yohe. "A Globally Coherent Fingerprint of Climate Change Impacts across Natural Systems." *Nature* 421 (2003): 37–42.
Paulson, Susan. "Degrowth: Culture, Power, and Change." *Journal of Political Ecology* 24 (2017): 425–48.
Pedersen, Poul. "Nature, Religion and Cultural Identity: The Religious Environmentalist Paradigm." In *Asian Perceptions of Nature: A Critical Approach*, ed. Ole Bruun and Arne Kalland, 258–76. London: Curzon, 1995.
Peterson, Hikaru Hanawa, Andrew Barkley, Adriana Chacon-Cascante, and Terry L. Kastens. "The Motivation for Organic Grain Farming in the United States: Profits, Lifestyle, or the Environment?" *Journal of Agricultural and Applied Economics* 44, no. 2 (May 2012): 137–55.
Pinchot, Gifford. *The Fight for Conservation*. New York: Doubleday, 1910.
Place, Elizabeth. "Land Use." In Kraybill, *Amish and the State*, 191–210.
Pope Francis. *Laudato si': On Care for Our Common Home*. Rome: Vatican Press, 2015.
Pritchard, James A. "A Landscape Transformed: Ecosystems and Natural Resources in the Midwest." In *Rural Midwest since World War II*, ed. J. L. Anderson. Dekalb: Northern Illinois University Press, 2014.
Quillin, Patrick. *The Wisdom of Amish Folk Medicine*. New Haven, MO: Leader, 1993.
Rametsteiner, Ewald, and Markku Simula. "Forest Certification—An Instrument to Promote Sustainable Forest Management?" *Journal of Environmental Management* 67 (2003): 87–98.
Rappaport, Roy A. *Ecology, Meaning, and Religion*. Berkeley, CA: North Atlantic Books, 1979.
Redekop, Calvin, ed. *Creation and the Environment: An Anabaptist Perspective on a Sustainable World*. Baltimore: Johns Hopkins University Press, 2000.
Reed, Robert, Anna Reed, Patrick McArdle, Michael Miller, Toni Pollin, Alan Shuldiner, Nanette Steinle, and Braxton Mitchell. "Vitamin and Supplement Use among Old Order Amish: Sex-Specific Prevalence and Associations with Use." *Journal of the Academy of Nutrition and Dietetics* 115, no. 3 (2015): 397–405.
Reiheld, Amelia T. "Donald G. Beam." In *The Amish of Holmes County*, ed. Jon Kinney, 241–49. Orrville, OH: Spectrum, 1996.

Reschly, Steven D. *The Amish on the Iowa Prairie, 1840–1910*. Baltimore: Johns Hopkins University Press, 2000.

______. "The Midwestern Amish since 1945." In *Rural Midwest since World War II*, ed. J. L. Anderson, 276–95. Dekalb: Northern Illinois University Press, 2014.

Ripple, William J., Christopher Wolf, Mauro Galetti, Thomas M. Newsome, Mohammed Alamgir, Eileen Crist, Mahmoud I. Mahmoud, and William F. Laurance. "World Scientists' Warning to Humanity: A Second Notice." *BioScience* 67, no. 12 (2017): 1026–28.

Ritzer, George. *The McDonaldization of Society*. 8th ed. Thousand Oaks, CA: Sage, 2014.

Robbins, Joel. "Properties of Nature, Properties of Culture: Ownership, Recognition, and the Politics of Nature in a Papuan New Guinea Society." In *Reimagining Political Ecology*, ed. Aletta Biersack and James B. Greenberg, 171–91. Durham, NC: Duke University Press, 2006.

Robbins, Paul. *Political Ecology*. 2nd ed. Hoboken, NJ: Wiley-Blackwell, 2011.

Robbins, Paul, and Julie Sharp. "The Lawn-Chemical Economy and Its Discontents." In Haenn, Wilk, and Harnish, *Environment in Anthropology*, 159–69.

Robinson, Gordon. *The Forest and the Trees: A Guide to Excellent Forestry*. Washington, DC: Island Press, 1988.

Ryan, John D., and Alan Thein Durning. *Stuff: The Secret Lives of Everyday Things*. Seattle: Northwest Environment Watch, Seattle, 1997.

Sanderson, Eric W., Malanding Jaiteh, Marc A. Levy, Kent H. Redford, Antoinette V. Wannebo, and Gillian Woolmer. "The Human Footprint and the Last of the Wild." *BioScience* 52 (2002): 891–903.

Scheffers, Brett R., Luc De Meester, Tom C. L. Bridge, Ary A. Hoffman, John M. Pandolfi, Richard T. Corlett, Stuart H. M. Butchart, Paul Pearce-Kelly, Kit M. Kovacs, David Dudgeon, Michela Pacifici, Carlo Rondinini, Wendy B. Foden, Tara G. Martin, Camilo Mora, David Bickford, and James E. M. Watson. "The Broad Footprint of Climate Change from Genes to Biomes to People." *Science* 354 (2016): aaf7671-1–aaf7671-11.

Schewe, Rebecca, and Caroline Brock. "Stewarding Dairy Herd Health and Antibiotic Use on U.S. Amish and Plain Mennonite Farms." *Journal of Rural Studies* 58 (2018): 1–11. doi: 10.1016/j.jrurstud.2017.12.023.

Schneider, F., G. Kallis, and J. Martinez-Alier. "Crisis or Opportunity? Economic Degrowth for Social Equity and Ecological Sustainability." *Journal of Cleaner Production* 18, no. 6 (2010): 511–18.

Schultz, P. Wesley, John B. Unipan, and Raymond J. Gamba. "Acculturation and Ecological Worldview among Latino Americans." *Journal of Environmental Education* 31 (2000): 22–27.

Serpell, James. *In the Company of Animals: A Study of Human-Animal Relationships*. New York: Cambridge University Press, 1986.

Sheridan, Thomas E. *Landscapes of Fraud: Mission Tumacacori, the Baca Float, and the Betrayal of the O'odham*. Tucson: University of Arizona Press, 2008.

Sherman, David M. *Tending Animals in the Global Village*. Baltimore: Lippincott, Williams & Wilkins, 2002.

Shetler, Paul. "Game Warden Leaves Us with an Impression." In *Legendary Adventures: Priceless Hunting Memories from Sportsmen Who Enjoy Sharing Their Experiences*, ed. Jonas M. Mast, 37–38. Sugarcreek, OH: Carlisle, 2007.

Shoreman-Ouimet, E., and Helen Kopnina. *Culture and Conservation: Beyond Anthropocentrism*. New York: Routledge, 2016.

Sleeth, Nancy. *Almost Amish: One Woman's Search for a Simpler, Slower Lifestyle*. New York: Tyndale House, 2012.

Smith, Eric Alden, and Mark Wishnie. "Conservation and Subsistence in Small-Scale Societies." *Annual Review of Anthropology* 29 (2000): 493–524.

Speth, James Gustave. *Red Sky at Morning: America and the Crisis of Global Environmentalism.* 2nd ed. New Haven, CT: Yale University Press, 2004.

Sponsel, Leslie E. *Spiritual Ecology: A Quiet Revolution.* Oxford: Praeger, 2012.

Stambaugh, Bruce. "Rare Birders Find Rare Birds." *Bird Watcher's Digest* 39, no. 3 (Jan./Feb. 2017): 47–49.

State of Ohio Environmental Protection Agency, Division of Surface Water. "Recreational Use Water Quality—Survey of the Sugar Creek Watershed, Holmes, Stark, Tuscarawas, and Wayne Counties, Ohio." 2005. http://epa.ohio.gov/portals/35/documents/2005FINALSugarCreekBactiReport.pdf

———. "Total Maximum Daily Loads for the Sugar Creek Basin: Final Report." 2002. http://www.epa.ohio.gov/portals/35/tmdl/SugarCreekTMDL_Final2002pdf.

Stein, Michelle M., Cara L. Hrusch, Justyna Gozdz, Catherine Igartua, Vadim Pivniouk, Sean E. Murray, Julie G. Ledford, Mauricius Marques dos Santos, Rebecca L. Anderson, Nervana Metwali, Julia W. Neilson, Raina M. Maier, Jack A. Gilbert, Mark Holbreich, Peter S. Thorne, Fernando D. Martinez, Erika von Mutius, Donata Vercelli, Carole Ober, and Anne I. Sperling. "Innate Immunity and Asthma Risk in Amish and Hutterite Farm Children." *New England Journal of Medicine* 375 (2016): 411–21.

Stevick, Richard. *Growing Up Amish.* 2nd ed. Baltimore: Johns Hopkins University Press, 2015.

Stinner, Deborah H., Maurizio G. Paoletti, and Ben R. Stinner. "In Search of Traditional Farm Wisdom for a More Sustainable Agriculture: A Study of Amish Farming and Society." *Agriculture, Ecosystems and Environment* 27 (Nov. 1989): 77–90.

Stoddart, Mark C. J., and Elahe Nezhadhossein. "Is Nature-Oriented Tourism a Pro-Environmental Practice? Examining Tourism-Environmental Alignments through Discourse Networks and Intersectoral Relationships." *Sociological Quarterly* 57, no. 3 (Summer 2016): 544–68.

Stone, Glenn Davis. "The Anthropology of Genetically Modified Crops." *Annual Review of Anthropology* 39 (2010): 381–400.

Sustainable Agriculture Research and Education (SARE). *The New American Farmer: Profiles of Agricultural Innovation.* 2nd ed. Beltsville, MD: SARE Outreach, 2005.

Taylor, Bron. *Dark Green Religion: Nature Spirituality and the Planetary Future.* Berkeley: University of California Press, 2010.

———. "Wilderness, Spirituality, and Biodiversity in North America: Tracing an Environmental History from Occidental Roots to Earth Day." In *Wilderness Mythologies: Wilderness in the History of Religions*, ed. Laura Feldt, 293–324. Berlin: DeGruyter, 2012.

Taylor, Bron, Gretel Van Wieren, and Bernard Daley Zaleha. "Lynn White Jr. and the Greening-of-Religion Hypothesis." *Conservation Biology* 30, no. 5 (2016): 1000–1009.

Taylor, Dorceta. *Toxic Communities: Environmental Racism, Industrial Pollution, and Residential Mobility.* New York: New York University Press, 2014.

Taylor, Peter J., and Frederick H. Buttel. "How Do We Know We Have Global Environmental Problems? Science and the Globalization of Environmental Discourse." In Haenn, Wilk, and Harnish, *Environment in Anthropology*, 202–13.

Tharp, Bruce M. "Valued Amish Possessions: Expanding Material Culture and Consumption." *Journal of American Culture* 30, no. 12 (Mar. 2007): 38–53.

Theodori, Gene L., A. E. Luloff, and Fern K. Willits. "The Association of Outdoor Recreation and Environmental Concern: Reexamining the Dunlap-Heffernan Thesis." *Rural Sociology* 63, no. 1 (1998): 94–108.

Trollinger, Susan L., and William Vance Trollinger, Jr. *Righting America at the Creation Museum.* Baltimore: Johns Hopkins University Press, 2016.

Ulrich-Schad, Jessica D., Caroline Brock, and Linda S. Prokopy. "A Comparison of Awareness, Attitudes, and Usage of Water Quality Conservation Practices Between Amish and non-Amish Farmers. *Society and Natural Resources* 30, no. 12 (2017): 1476–90. doi: 10.1080/08941920.2017.1364457.

United States Department of Agriculture, National Agricultural Statistics Service. "Organic Farming: Results from the 2014 Organic Farming Survey." *Agricultural Census Highlights* 29, no. 12 (Sept. 2015).

Urry, John. *The Tourist Gaze: Leisure and Travel in Contemporary Societies*. Thousand Oaks, CA: Sage, 1990.

Vincent, Carol Hardy, Laura A. Hanson, and Carla N. Argueta. *Federal Land Ownership: Overview and Data*. Congressional Research Service Report 7-5700. Washington, DC: Congressional Research Service, 3 Mar. 2017. https://fas.org/sgp/crs/misc/R42346.pdf.

Vitousek, Peter M., Harold A. Mooney, Jane Lubchenco, and Jerry M. Melillo. "Human Domination of Earth's Ecosystems." *Science* 277 (25 July 1997): 494–99.

Vonk, Martine. *Sustainability and Quality of Life: A Study on the Religious Worldviews, Values and Environmental Impact of Amish, Hutterite, Franciscan and Benedictine Communities*. Amsterdam: Buijten & Schipperheijn Motief, 2011.

Wackernagel, Mathis, and William Rees. *Our Ecological Footprint: Reducing Human Impact on the Earth*. Gabriola Island, BC: New Society, 1998.

Walker, Peter A. "Political Ecology: Where Is the Ecology?" *Progress in Human Geography* 29, no. 1 (2005): 73–82.

Waters, Colin N., Jan Zalasiewicz, Colin Summerhayes, Anthony D. Barnosky, Clément Poirier, Agnieszka Galuszka, Alejandro Cearreta, Matt Edgeworth, Erle C. Ellis, Michael Ellis, Catherine Jeandel, Reinhold Leinfelder, J. R. McNeill, Daniel deB. Richter, Will Steffen, James Syvitski, Davor Vidas, Michael Wagreich, Mark Williams, An Zhisheng, Jacques Grinevald, Eric Odada, Naomi Oreskes, and Alexander P. Wolfe. "The Anthropocene is Functionally and Stratigraphically Distinct from the Holocene." *Science* 351, no. 6269 (Jan. 2016). doi: 10.1126/science.aad2622.

Watts, Michael, and Richard Peet. "Liberating Political Ecology." In *Liberating Ecologies: Environment, Development, Social Movements*, ed. Peet and Watts, 3–47. 2nd ed. New York: Routledge, 2004.

Weaver, Mark R., Richard H. Moore, and Jason Shaw Parker. "A Farmer Learning Circle: The Sugar Creek Partners, Ohio." In *Pathways for Getting to Better Water Quality: The Citizen Effect*, ed. Lois Wright Morton and Susan S. Brown, 197–212. New York: Springer, 2011.

Weaver-Zercher, David L. *The Amish in the American Imagination*. Baltimore: Johns Hopkins University Press, 2001.

Weaver-Zercher, Valerie. *Thrill of the Chaste: The Allure of Amish Romance Novels*. Baltimore: Johns Hopkins University Press, 2013.

Weber, Christopher L., and H. Scott Matthews. "Food-Miles and the Relative Climate Impacts of Food Choices in the United States." *Environmental Science and Technology* 42 (2008): 3508–13.

Welk-Joerger, Nicole. "Religion, Science, and Diverse Farming Methods among the Amish in Pennsylvania." Paper presented at the Continuity and Change: 50 Years of Amish Society Conference, Elizabethtown College, PA, 10 June 2016.

Wenger, Olivia K., Mark D. McManus, John R. Bower, and Diane L. Langkamp. "Underimmunization in Ohio's Amish: Parental Fears Are a Greater Obstacle Than Access to Care." *Pediatrics* 128, no. 1 (July 2011): 79–85.

Wengerd, Eli. "The Joy of Taking the Next Generation Hunting." In *Legendary Adventures: Priceless Hunting Memories from Sportsmen Who Enjoy Sharing Their Experiences*, ed. Jonas M. Mast, 165–66. Walnut Creek, OH: Carlisle, 2007.

Wesner, Erik. *Success Made Simple: An Inside Look at Why Amish Businesses Thrive*. New York: Jossey-Bass, 2010.

Wexler, Jay. *When God Isn't Green: A World-Wide Journey to Places Where Religious Practice and Environmentalism Collide.* Boston: Beacon, 2016.

White, Lynn, Jr. "The Historical Roots of Our Ecologic Crisis." *Science* 155, no. 3767 (Mar. 1967): 1203–7.

White, Richard. *Middle Ground: Indians, Empires, and Republics in the Great Lakes Region, 1650–1815.* New York: Cambridge University Press, 1991.

Whitney, Gordon G. *From Coastal Wilderness to Fruited Plain: A History of Environmental Change in Temperate North America, 1500 to the Present.* Cambridge: Cambridge University Press, 1994.

Widmann, Richard H. "Forests of Ohio, 2015." Resource Update FS-97. Newtown Square, PA: USDA, Forest Service, Northern Research Station, 2016.

Widner, David E. *Old Order Amish Beliefs About Environmental Protection and the Use of Best Management Practices in the Sugar Creek Watershed.* MA thesis, Kent State University, 2010.

Wilson, E. O. "The Little Things That Run the World (The Importance and Conservation of Invertebrates)." *Conservation Biology* 1, no. 4 (1987): 344–46.

World Commission on Environment and Development of the United Nations. *Our Common Future.* The Brundtland Report. Oxford: Oxford University Press, 1987.

Yoder, Rhonda Lou. "Amish Agriculture in Iowa: A Preliminary Investigation." MA thesis, Iowa State University, 1990.

Young Center Books in Anabaptist & Pietist Studies

Simon J. Bronner and Joshua R. Brown, eds., *Pennsylvania Germans: An Interpretive Encyclopedia*

James A. Cates, *Serving the Amish: A Cultural Guide for Professionals*

D. Rose Elder, *Why the Amish Sing: Songs of Solidarity and Identity*

Brian Froese, *California Mennonites*

Felipe Hinojosa, *Latino Mennonites: Civil Rights, Faith, and Evangelical Culture*

Charles E. Hurst and David L. McConnell, *An Amish Paradox: Diversity and Change in the World's Largest Amish Community*

Rod Janzen and Max Stanton, *The Hutterites in North America*

Karen M. Johnson-Weiner, *Train Up a Child: Old Order Amish and Mennonite Schools*

Peter J. Klassen, *Mennonites in Early Modern Poland and Prussia*

James O. Lehman and Steven M. Nolt, *Mennonites, Amish, and the American Civil War*

Mark L. Louden, *Pennsylvania Dutch: The Story of an American Language*

David L. McConnell and Marilyn D. Loveless, *Nature and the Environment in Amish Life*

Steven M. Nolt, *The Amish: A Concise Introduction*

Steven M. Nolt and Thomas J. Meyers, *Plain Diversity: Amish Cultures and Identities*

Douglas H. Shantz, *An Introduction to German Pietism: Protestant Renewal at the Dawn of Modern Europe*

Tobin Miller Shearer, *Daily Demonstrators: The Civil Rights Movement in Mennonite Homes and Sanctuaries*

Janneken Smucker, *Amish Quilts: Crafting an American Icon*

Richard A. Stevick, *Growing Up Amish: The Rumspringa Years* (second edition)

Duane C. S. Stoltzfus, *Pacifists in Chains: The Persecution of Hutterites during the Great War*

Susan L. Trollinger, *Selling the Amish: The Tourism of Nostalgia*

Diane Zimmerman Umble and David L. Weaver-Zercher, eds., *The Amish and the Media*

David L. Weaver-Zercher, *Martyrs Mirror: A Social History*

Valerie Weaver-Zercher, *Thrill of the Chaste: The Allure of Amish Romance Novels*